SpringerBriefs in Ethics

Springer Briefs in Ethics envisions a series of short publications in areas such as business ethics, bioethics, science and engineering ethics, food and agricultural ethics, environmental ethics, human rights and the like. The intention is to present concise summaries of cutting-edge research and practical applications across a wide spectrum.

Springer Briefs in Ethics are seen as complementing monographs and journal articles with compact volumes of 50 to 125 pages, covering a wide range of content from professional to academic. Typical topics might include:

- Timely reports on state-of-the art analytical techniques
- A bridge between new research results, as published in journal articles, and a contextual literature review
- A snapshot of a hot or emerging topic
- In-depth case studies or clinical examples
- Presentations of core concepts that students must understand in order to make independent contributions

More information about this series at http://www.springer.com/series/10184

Bashir Jiwani

Good Organizational Decisions

Ethical Decision-Making Toolkit for Leaders and Policy Makers

 Springer

Bashir Jiwani
Fraser Health Ethics Services
and Diversity Services
Burnaby, BC, Canada

ISSN 2211-8101   ISSN 2211-811X   (electronic)
SpringerBriefs in Ethics
ISBN 978-3-030-33400-0   ISBN 978-3-030-33401-7   (eBook)
https://doi.org/10.1007/978-3-030-33401-7

This Springer imprint is published by the registered company Springer Nature Switzerland AG
The registered company address is: Gewerbestrasse 11, 6330 Cham, Switzerland

Introduction

If consulting *Good Organizational Decisions: Ethical Decision-Making Toolkit for Leaders and Policy Makers* (I'll refer to this as *The Toolkit* going forward), you are likely already interested in, if not motivated to use, a structured ethics-based method for decision-making. *The Toolkit* describes a step-by-step guide for one such approach. It is the product of research[1] that explored two key questions:

(1) What determines the right, or ethically justified, answer to system-level decisions made by public or civil society institutions?
(2) What would assist leaders in making more ethically justified system-level decisions?[2]

The approach described in *The Toolkit* is nuanced and requires care in application, as well as time, energy, and resources to employ. In this introduction, I outline the values behind the method and clarify some possible benefits. I then provide an overview of the various phases included in the approach. My goal is to give readers confidence the approach is worth the investment and to provide language to help explain the approach to key partners.

System level issues, problems, questions, and decisions are inescapably about values. A system-level decision is always a reflection of what matters—what decision-makers value and value most. In the health care context for example, a hospital might make the system-level decision to invest in training for social workers to improve their ability to have good end-of-life discussions with patients and families. Hospital leaders may make this choice because they value supporting patients and families through the end of life journey. Or, because funders have

---

[1]Jiwani, B. (2008). *Ethically Justified System-Level Decisions in Health Care: Toward a Decision Support Workbook for Healthcare Leaders* (Unpublished doctoral dissertation). University of Alberta, Canada.

[2]The system-level method presented here is part of a suite of tools including *Clinical Ethics Consultation: A Practical Guide* and *Clinical Ethics Consultation Toolkit* both published by Springer (Jiwani, 2017). Additional resources are also available on the supporting website *incorporatingethics.ca*.

requested this and hospital leadership values pleasing funders. Or, the hospital might train social workers because this commitment is in their contract and hospital leadership values honouring contracts. In the education context, leaders of a school might choose to begin a school lunch program. They may choose this to deter children from leaving school grounds—perhaps for reasons of safety or maximizing time for learning. Or they may choose this because children are not being sufficiently nourished at home—in which case it might be student health and wellbeing, equity and social justice, or effectiveness in learning that education leaders value.

There may be many reasons why we make a choice. Ultimately, a decision is made to gain one or multiple things of value. The benefit might be direct (happier, more skilled social workers; students present onsite) or indirect (supported patients and families, happier health care teams; students better able to learn effectively), but the benefit is something we value.

System-level decisions affect how systems are organized, governed and operated, as well as program goals, access and staffing. System-level decisions often concern the allocation of resources—from materials and supplies to access to facilities and experts. Who will benefit from resources and who will suffer the harms of going without? Who will be charged with making these decisions? What process will be used in the decision-making? All of these are system-level decisions. At the heart of each choice—or set of choices—is what the organization values.

Good decisions are ethically justified decisions. Leaders generally seek to make good decisions. A decision is good if it effectively solves a problem; is based on relevant, well-justified values; is supported by facts; and is developed in partnership with those affected to live up to what matters to them, without anyone being required to involuntarily change their core commitments. A good decision is an ethically justified decision, and the degree of a decision's ethical justification depends on how well these conditions are met.

The method described in *The Toolkit* is concerned with both content and process. Ethical justification requires getting the values—and the corresponding facts—right. But how we get there matters. On this approach, an ethically justified decision emerges through a collaborative process of moral reasoning which forges a common understanding of beliefs about values and facts.

There are several benefits to this method, including: solutions that are better justified; decisions that more effectively identify and address the central problem; greater compliance with decisions by those affected; and decreased moral distress on the part of those working within the system.

Careful and systematic attention to values in a decision process is required for individual and organizational integrity. Explicit attention to the values that a system-level decision should be based on leads to policies, practice standards and guidelines, strategies, and outlooks that are more consistent with the values of the organization and the individuals that comprise it. Taking the time to examine what we should hold important (values) and what we think is true about the world (facts) related to these values leads to making decisions, taking actions, and having attitudes that are more consistent with our respective views of the world.

Living with integrity is at the core of wellbeing for people and organizations. At an individual level, compromises to our abilities to understand and act on what matters to us leaves us feeling confused and uncertain, pursuing mixed up courses of action. When we have clarity about what commitments matter, but we are prevented from living these out, we experience being torn apart, feeling pain and suffering at the feared or actual loss of what we care about.

At the organization level, an unclear sense of what matters or a disconnect between key commitments and organizational policies and practices can lead to team members feeling at odds with each other, deviation from accepted standards and associated harms, and inconsistent practices.

Conversation spaces that enable pluralistic deliberation improve the quality of decision-making. Today, most of us live with, work alongside, and serve people who are different than us, and hold different worldviews. In part, this is why system-level decision-making is so difficult. If we lived in homogenous communities, the values guiding decision-making would be ready at hand, because shared values would be implicit. But, oftentimes, we are different than our neighbours, the people we encounter at work, and those whom we meet at the community center.

While human beings have differing beliefs about what matters, they are also able to grow and change commitments as their understanding of situations and each other deepens. More importantly, an individual's identity (including self-understanding and value and belief commitments) is shaped through the process of participating in community with other people.

Pluralism is a response to diversity that sees difference as a blessing, albeit a challenging one. In conversation spaces that are pluralist, members of the group learn about others' ideas, respond to these ideas, and have their own perspectives understood and responded to in turn.

Such conversation spaces are based on values of equality and inclusiveness. In an ideal deliberation process, all would have equal standing at each phase (deciding the agenda, proposing and debating options, and making decisions or recommendations). In this space, the strength of reasons would guide decision-making, not power in the form of economic or political standing.

Pluralistic conversations allow people to understand different perspectives and to shift their own in consideration of the group. Participants are more likely to accept the decisions that come out of pluralist conversation—sometimes because they feel included in the process, sometimes because they agree with the reasoning, sometimes because the discussion has given them a broader understanding of the issues involved, and sometimes because they can foresee these reasons supporting their own positions in future and more important decisions.

A decision team's structure, mandate, and methods matter greatly. Traditional approaches of ethics analysis usually involve identifying what principles should inform a decision and assessing whether or how well a decision lives up to these principles. These methods, focusing on a decision itself, leaves many gaps. This is because the beliefs about values and facts that eventually inform a decision are shaped at multiple levels, beginning with who is developing the decision on how the issues are analyzed.

For the approach used to make systems-level decisions to enable justified values and solid facts to emerge and inform choices made it should explicitly consider:

- What is the problem being addressed?
- Is the desired deliverable a decision, policy, or guideline?
- Who is leading the decision team?
- Who is on the decision team?
- How will the members of the decision team relate to one another?
- What kinds of reasons will count as good ones?
- What standards of evidence will be required before information is accepted as fact?
- Who is the decision team accountable to?

The decision-making group will have to be clear on its authority to make decisions. They will need to know whether their decisions are binding, or whether they are recommendations. If these are recommendations, to whom are they made, and what expectation does the team have that the recommendations will be taken seriously? *The Toolkit* expands on these and other questions.

A facilitator trained in ethics can help deepen the values analysis. This Toolkit is designed for any decision-makers to use. It is not meant to require a facilitator or an ethicist to use. The reasons for this are that not every organization has an ethicist available to assist and because, in my opinion, leaders need to develop the skills to undertake the kind of moral analysis described in the Toolkit.

Ethicists have studied moral philosophy, will be familiar with social norms, and have been trained to identify and evaluate assumptions and understand the deeper commitments that lay beneath superficial value descriptions. When the process is undertaken with the participation or leadership of an ethicist, this can assist the group to ensure it takes into consideration relevant value commitments in society, reflect on other perspectives that might not be obvious to the group to think about, and explain the deeper principles and moral traditions upon which common values are based. For these reasons it can be very useful to include an ethicist in the process or to have a facilitator with ethics training guide the process.

Procedural consensus should be the standard for making decisions and resolving disagreement. A common challenge faced by decision teams is that the members disagree on various issues. Usually when this happens, decisions are made by the last, loudest, or most powerful voice at the table. But these voices may not reflect values that can stand up to close scrutiny. Establishing a justified process of how decisions are made means asking how disagreement will be resolved among those on the decision team.

If genuinely respecting and responding to all different perspectives and hearing and considering a variety of viewpoints are important to the team (i.e., they are the team's process values), then the team will likely adopt a consensus model of resolving disagreement. If, on the other hand, deciding quickly and deferring to those in authority are the team's values, then brief discussion followed by a vote or decision by the most powerful group member will be the model for resolving disagreement.

Coming to complete, substantive agreement on the best answer to a difficult question is unlikely to occur in a short span of time. However, procedural consensus is possible. This occurs when all members of a group agree that the discussion has been fair, all perspectives have been solicited, and the reasons behind perspectives have been understood and meaningfully engaged. So even though there may be disagreement within the group about the decision or recommendation, there is consensus that the process has been fair. When procedural consensus has been reached and time has run out, the group advances and stands behind the recommendation of the majority, based on the reasons discussed and keeping in mind the counter-concerns raised by the dissenting group members. The dissent and the reasons for the minority view can also be captured to support ongoing analysis of the issue over time.

Consultation with those affected is key. When it comes to health care, system-level decisions have an enormous impact on individuals and the community at large. Yet there is almost no community consultation involved with most system-level decision-making. Some would suggest that this failure to engage the community ignores democratic values that are crucially important in Western, liberal society and therefore diminishes justification of these decisions. The approach to system-level decision-making described in *The Toolkit* takes this concern seriously and argues that decision-makers need to think carefully about community consultation in their decision-making process.

One challenge in health system decision-making is that we are not used to thinking rigorously about how to consult with the community. The fact that it is so complicated to determine appropriate representation, the extent of citizen involvement, how to handle interest groups, and what to do in the face of disagreement does not help. Another challenge that faces decision-makers is that even after some form of consultation is deemed necessary, specialized knowledge and skills are required to undertake community consultation.

Regardless of these challenges, decision-makers have a responsibility to build and engage systems of community consultation, if they are to make democratically legitimate decisions.

Integrity requires living our values, not just naming them—so implementation is crucial. Making decisions is one thing. Carrying out the change management required to bring decisions alive is quite another. Not implementing the emerging plan that would allow the organization to live up to the values determined to be the most well-justified in the context may provide pretty window-dressing, but would leave unaddressed the rot found in the structure. Decision leaders need to ensure an implementation plan is developed and executed. They also need to anticipate and support decision follow-up with:

- Education plans to train those implementing decisions.
- Communication plans to share relevant information about decisions.
- Sustainability plans to ensure the long-term implementation of well-considered and justified decisions.

- Downstream support plans to assist those put in morally compromising situations as a result of upstream decisions.
- Evaluation and review plans to ensure the actual consequences of a decision match those anticipated, and to alter direction in light of any changes in facts or values that may arise.

The approach described in The Toolkit can be divided into five key phases. Although transitions between these are fluid, it can be helpful to distinguish the approach into the following stages.

Phase 1: Establishment of the decision team. *The Toolkit* calls for the creation of a shared work team that will be responsible for decision-making and establishing the parameters of the team's work. The team would agree about how they will work together, what they will produce (a decision or recommendation) and to whom this decision or recommendation will be submitted.

Phase 2: Development of a preliminary decision through moral analysis. The team undertakes a systematic process of values-guided critical thinking and solution-finding to arrive at a preliminary decision. In this phase a consultation plan is also developed. The team would analyze the relevant values and facts at stake in the situation, prioritize the values, brainstorm possible solutions, test these against the prioritized values and arrive at a preliminary decision. This preliminary decision would be articulated together with the rationale on which it is based.

Phase 3: Consultation. Appropriate consultation is undertaken and a document consolidating the feedback received is prepared. The team would identify (or, identifies) relevant experts and those affected by the decision. The decision team would determine how best to share the team's thinking and solicit feedback from these individuals and groups about what should matter most, what the relevant facts on the ground are and what the best decision might be.

Phase 4: Development of a final decision. In this phase the shared work team carefully assesses the feedback, revises the decision accordingly and submits its decision together with its rationale to the appropriate authority. The decision team also develops a response document to provide to those engaged in consultation.

Phase 5: Implementation Planning. The decision team develops an implementation plan, together with detailed plans for communication, education, downstream support and evaluation. The decision team would transfer leadership of the decision to the appropriate body for enacting the implementation.

Attending to these lessons from experience can assist with implementation of the process. The approach described in *The Toolkit* is the foundation for a System-Level Ethics Consultation Service offered by Fraser Health Ethics and Diversity Services. The approach has been used for over a decade on dozens of system-level issues. It has been employed to establish broad response strategies, such as for dealing with the threat of Ebola virus disease, changes to legislation for medical assistance in dying, and the place of medical and non-medical cannabis in health care. It has been used to develop frameworks for the allocation of drugs and medical supplies in short supply. It has also been used to support teams to develop strategies to respond to issues ranging from dealing with moral distress to assisting with complex changes to program structures and practices. This experience has provided important lessons about facilitating ethics decision processes. Three important lessons are as follows.

The more complex the key question, the longer the process will take. Some key questions involve numerous sub-questions. For example, questions of resource allocation usually include three different questions: what criteria should be used to discriminate between alternative candidates for the resources, who should be empowered to use the criteria to make the decisions, and through what process should options be raised, decisions analyzed and made, and decisions be implemented? Tackling problems with multiple sub-questions will take longer because the values that will need to be brainstormed, prioritized and justified for each of these questions is slightly different. Recognizing the complexity of the key question and setting appropriate expectations at the outset will be important for effectively solving the problem in a satisfying manner for those involved.

Illustrating the value of the process by using one exercise can help to build the necessary trust to move forward with an overall process. The approach in The Toolkit requires time, energy, commitment and good faith participation. These are precious resources and so system leaders will need to place trust in the process and its facilitators that the investment will be worthwhile. One way of establishing the credibility of the facilitator and process is leading a single workshop exercise focusing on one of the steps in *The Toolkit*. For example, a team struggling with a system-level issue could be led through the exercise of simply brainstorming what each team member thinks is important that the solution should live up to. As people share their thoughts, the facilitator could distinguish between the facts, values and dimensions of the solution that will inevitably emerge in such a discussion. Helping participants to see these different dimensions and recognize that dealing with each systematically, as The Toolkit enables, will be beneficial (indeed necessary) in making a good decision, can go a long way in earning participants' trust of the process. Such an exercise can be aided by using the online version of the analysis tool available at http://incorporatingethics.ca/public_files/exercise/ethics_analysis.htm.

Careful documentation and transparent sharing of the workbook with the shared work team will inspire confidence, ensure key discussion points are captured accurately and enable easier values analysis. The steps in the Toolkit each come with corresponding worksheets. These worksheets can be combined into a workbook (a template is available online at http://incorporatingethics.ca/good-decisions/).

For the first parts of the process where the team is brainstorming facts and values, capturing these on a flipchart or whiteboard so everyone can see what is being written will help create common purpose and understanding. Having someone participate in the meetings to capture the discussion directly into the worksheets, especially during later phases of the process, will save time and allow easy distribution of the group's thinking to shared work team members for review and correction. At step 4, when the group prioritizes values, having the values listed in the table with their scores will make it easier for the facilitators to arrange the values statements according to theme and priority in different formats.

I hope you will give the approach a try and that it serves you well. The use of
this approach has enabled high-quality policies, guidelines, and strategies to
emerge; ongoing interest and satisfaction has been shown by policy leaders and
those who have used this framework. I am very hopeful that *The Toolkit* can and
will continue to enable organizations, the people who make up these bodies, and
perhaps most importantly, the individuals and communities served by these bodies
to live more integrated, more just, happier, and more peaceful lives.

Readers interested in the model but who have questions about how to use it are
welcome to reach out to me or the team at Fraser Health Ethics and Diversity
Services at ethics.services@fraserhealth.ca.

I would like to especially acknowledge my colleagues Susan Rink and Duncan
Steele for their incredible championship and support of this approach. I also express
my gratitude to Fraser Health, and the now passed on Provincial Health Ethics
Network of Alberta for their support and commitment to making and supporting
ethically justified system-level decisions.

Good Organizational Decisions:

Ethical Decision-Making Toolkit for Leaders and Policy Makers.

Bashir Jiwani, PhD

Welcome to this Resource

SYSTEM-LEVEL DECISIONS

The term "system-level", as used throughout this document, refers to any issue or decision that concerns more than one or two specific cases or situations, or that affects others beyond the particular people in a given scenario.

In healthcare, system-level issues can concern clinical practice – for example, a team's approach to patients who don't follow through on their self-care plans. They can also involve organizational issues - for instance, how should housekeeping services be set up?

Outside of healthcare, system-level issues include, for example, human resources issues – such as professional development policies for organizational staff who have a wide range of professional responsibilities.

Dear Reader,

Welcome to this toolkit and process guide for making system-level decisions.

One of the services we provide at Fraser Health Ethics and Diversity Services (FHEDS) is ethics support for addressing system-level issues within the health region. We do this to help the organization deal with specific, challenging problems or questions and to build capacity to undertake ethics-based decision-making. The consultation process includes active facilitation through a structured decision process. This toolkit is a tool to help guide the process.

The toolkit is meant to aid others interested in exploring this approach. My sincere hope is that individuals and teams will find this toolkit useful in making their own system-level decisions.

The process described here is based on my research exploring two related questions: **What makes a given decision ethically justified?** And, **how do we assist leaders working on system-level issues to make better decisions?** A summary of the argument behind this approach is provided as an appendix at the end of the book.

We have used the approach to help organizations establish frameworks for allocating resources, determine the programs and services they will offer, develop staff support strategies through organizational change and changes in practices, put in place governance structures, to name a few. While our work has been predominantly in health care, we have also used the methods to support civil society organizations.

This book is part of a broader suite of decision-making resources, including an introduction and toolkit for clinical ethics consultation. Information about these resources is available at incorporatingethics.ca.

I warmly welcome any feedback you might have on the toolkit. If you have comments about the approach, questions about its method or how to use it, or wish to share your experience with it, I would be delighted to receive your email at bashir@bjei.ca. You may also reach out to the FHEDS team at ethics.services@fraserhealth.ca.

With good wishes,

Bashir Jiwani

Table of Contents

IF INTEGRITY IS ABOUT WALKING OUR TALK AND IF WE'RE NOT CLEAR ON WHAT OUR TALK IS, THEN HOW CAN WE WALK IT?

In a Nutshell

Integrity matters - for individuals and groups of people. When integrity is compromised, basic health and wellbeing and peace of mind for individuals are disturbed. And systems and communities experience inconsistency and unfairness, waste and vulnerability.

Living with integrity requires being in community with others. Trust is required for community to happen. Respectful engagement is the most effective way to build trust.

A good decision is a) based on a well-considered, justified understanding of what's important (values) and the best information (facts) available; and b) implemented in a manner that brings this thinking to life.

A transparent, inclusive, respectful, dialogical process that features consultation with experts and those affected is the best way to arrive at guiding values and facts.

Bringing a decision to life requires a sound execution and follow- up plan that includes careful attention to implementation, education, sustainability, evaluation, and downstream ethics support.

In many contexts, decisions are made without explicit attention to these key elements. This can lead to various types of problems, from alienation and disengagement to moral distress and inefficiency (as issues remain unresolved).

The approach in this book seeks to assist decision leaders or teams to realize the benefits of quality, legitimacy, and compliance - and most important, greater integrity for all concerned.

How to Use this Resource

ADDITIONAL RESOURCES ONLINE

This Toolkit is accompanied by a suite of resources available online at http://incorporatingethics.ca/

The accompanying tools include...

- Sample completed worksheets for the various steps

- A version of the toolkit, include a table for analyzing options against values

- More detailed descriptions of various aspects of moral deliberation and ethical decision-making.

This toolkit is meant for any individual or team that is looking for support in making an ethically-justified system-level decision – that is, a decision that will impact people in more than one or two individual situations.

To make the most of this toolkit:

- Read through it carefully (note what each step requires and the tips for success)
- Confirm a commitment to proceed and discuss possible key questions the group might work on (see Step 2)
- Consider using an external facilitator to guide the team through the discussion or assign this role to a member of the team
- Schedule three, two and a half-hour meetings to begin to work through the process in the context of your issue
- Ask members of the decision team to review steps 1 through 3 in advance of the first meeting
- Prepare a soft copy of the toolkit to enter the data from the meetings for circulation to the team for review before the next/second meeting. A soft copy is available at incorporatingethics.ca
- Review the Facilitation Techniques found on page 12 paying special attention to the distinction between facts, values and emotions
- Approach the exercise with a spirit of open-mindedness and flexibility. While the process looks linear the conversation will likely weave in and out through the various steps - and this is ok!

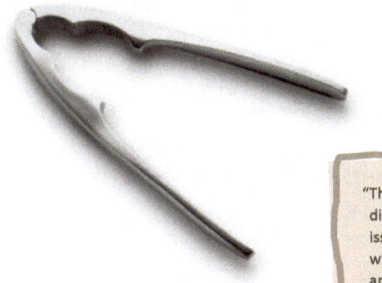

"The system level decision process has allowed for a variety of voices and disciplines to come together around some very complex and ethically challenging issues and decisions, and yet be able to methodically work through the process with clarity, thereby allowing for not only sound decisions, but also a consultative approach to implementing the decision. This methodology can be transformative to organizations and teams if applied well, with sound system level decisions being made based on a set of values that the organization or team feels is central to them. I wish we were able to make this a part of every team's functioning."

Dr. Mohamud Karim
Program Medical Director, Fraser Health Renal Program

What's Different About this Approach

Traditionally, when people think of ethics in the context of policy, what comes to mind are relevant key principles. Ethics analysis then explores what these principles mean and how they are balanced in the policy.

The conventional approach only considers formal policies as worthy of ethics analysis. This method of analysis leaves key ethically salient dimensions of the system-level process unaddressed.

Our approach prioritizes people, values and facts, careful consultation, clear communication at every step and an intentional implementation method.

Decision teams

WHAT MATTERS

Who is involved? How will disagreement be resolved? How open are we to different perspectives? The way a team communicates determine the values inherent in the solution. Process decisions should be made thoughtfully and intentionally.

CURRENT PRACTICE

A team is struck and immediately begins developing a solution. This usually begins by surveying literature and practice at other organizations. Little or no attention is paid to group mandate or dynamics.

THE INNOVATION

Our approach begins with explicit discussion of what the group's mandate is, who should be involved, and how the team will work.

Analysis of values and facts

WHAT MATTERS

Values and facts are different types of concepts. These different types of beliefs should be distinguished and analyzed separately.

CURRENT PRACTICE

Teams rarely separate the facts and values implicit in preferred positions. These are usually muddled together in defense of a preferred solution, making it hard to assess the quality of an argument or solution.

THE INNOVATION

Facts and values are explicitly treated as distinct categories. Each is addressed in a separate step and discussed in an organized and careful way.

Consultation with experts

WHAT MATTERS
Good decisions depend on good information. Relevant information includes not only technical facts, but also facts about the intended purpose or ends. To get this, the most knowledgeable sources should be consulted.

CURRENT PRACTICE
Technical information and practices at other organizations are often compiled well, but in an unsystematic way. Information about what matters to the experts involved and the values implicit in preferred solutions are often left unexplored.

THE INNOVATION
A separate worksheet is provided to help think through which physicians, staff and other experts should be consulted and how this can be done.

Consultation with those affected

WHAT MATTERS
By definition, system-level decisions affect others beyond the decision-makers. These individuals will have important knowledge about the context and will have things that matter to them that they will want to know have been considered in the decision-making. Their views should inform decisions.

For example, in healthcare, patients and family members often have the most at stake in system-level decisions. They also have important knowledge about the impact of clinical system-decisions.

CURRENT PRACTICE
Patients and families are rarely consulted in system-level decisions.

THE INNOVATION
A separate worksheet is provided to help think through which patients and loved ones should be consulted and how this can be done.

Consultation with the public

WHAT MATTERS
We share a commitment to ideals of democracy in Canada. This means that the values of those impacted by decisions should influence the decisions.

CURRENT PRACTICE
This is simply not considered in most system-level decision approaches.

THE INNOVATION
Separate tools are provided to help think through what sub-groups of the public should be consulted and how this can be done.

....

Decision rationale and justification

WHAT MATTERS
For people to be convinced that a decision or solution is the best course of action it would help for them to know the issue has been carefully thought through and for them to be able to assess the values and facts analysis. They should be provided decision rationale and justification.

CURRENT PRACTICE
Decisions tend to be written with incomplete descriptions of the context against which they are made. Often justifications incompletely articulate the values that the decisions live up to, and remain silent on values the decision sacrifices. The balancing of values is not discussed.

THE INNOVATION
Our approach supports the careful articulation of a preliminary decision before consultation and then again after a final decision is reached. This articulation includes a summary of important factual assumptions, important value trade-offs being made, and an explanation of why these are seen as defensible.

....

Decision follow-up plans

WHAT MATTERS
A decision itself won't solve the problem. The decision has to be implemented and supported. Implementation and follow up should be treated with care and be part of the decision process.

CURRENT PRACTICE
Decisions are made and then unevenly followed up. Practice standards are often supported by a short-term educational push, while policy decisions are often not followed up.

THE INNOVATION
In the process separate steps create space to discuss communication, education, sustainability, downstream ethics support, and evaluation plans.

Frequently Asked Questions

PATIENTS, RESIDENTS
AND CLIENTS

The words "patient" and "patients"
used in the text are meant to refer
to patients, residents and clients –
those receiving care or services in
the healthcare system.

Q Is this guide only appropriate for the healthcare context?

A The approach and process can be equally useful in a variety of situations –
really anywhere that system-level decisions are made.

Q What healthcare context is this intended for?

A The tool can be used in any healthcare context, from public health and
home care to long-term and acute care settings.

Q In what contexts has this approach been used?

A Within healthcare, it has been used in a variety of programs from mental
health & addictions and public health & primary care, to nephrology
and maternal & child health. Outside healthcare the process has been
used in private enterprise and civil society arenas to guide questions
concerning organizational structure, organizational core values, facility
use guidelines, and resource allocation criteria. For example, the method
has been used to determine how sacred spaces in religious buildings
should be used and how resources should be allocated in public service.

Q How quickly can we do this? How much time will this take?

A This will depend on the complexity of the issue and the comfort with the
process. For a key question that does not have multiple sub-questions, it
will likely take two meetings of between two and three hours each to get
to a preliminary decision. Three more meetings of the same duration, with
engagement and communication along the way, will probably be required
before the team will be ready for implementing the decision. (See the
suggested schedule for a sense of what is involved.)

Q Must we use the entire process?

A Using the process up to Step 7 may still offer a team improvement
from most existing methods of decision-making. Engagement (Step 8)
is necessary to maximally achieve the benefits of legitimacy and
compliance. The follow- up steps (especially Steps 11-14) are
necessary to achieve the benefits of effectiveness, especially in the
middle and long term.

Q When should teams use this tool?

A A team might be well-advised to use the tool first on a very important
policy decision that must be made but for which a time-horizon of a few
months is available. This will allow the group to go through the process
carefully and learn without feeling rushed. This first iteration will help
teams decide whether and how the process might become part of their
standard procedure.

....

Q What specific types of decisions have past teams made using this process?

A The approach has been used to assist teams to determine criteria and processes for allocating resources, to determine practice guidelines for responding to challenging clinical issues, and to assist newly formed programs to determine how they can best work together.

....

Q Is this a linear process?

A Although the process looks linear, there is much interplay between the steps. For example, it may not be until part way through the process that the key question becomes confirmed. And conversation about values may lead to the identification of new facts or missing information. While it is important that the process be complete, each step need not be completely finished before moving on to the next. (Indeed, sometimes it makes sense to switch certain steps around.)

....

Q Who should be involved in the process and at what point? Whose values should determine the answer to the question?

A The underlying principle of the approach is that if someone is impacted by a decision or has information that is relevant to the decision, that person should be able to contribute their wisdom. Determining the stage at which someone can share their perspective will require balancing efficiency and quality. Options include: being consulted on the decision team, evaluating the decision team's initial decision, being made aware of the purpose and progress of the work through the communication process, receiving education and support in the implementation stage, and participating in the decision's implementation and follow up.

....

Q Will using this process guarantee a good outcome?

A No approach to decision-making can overcome uncertainty. Despite the best and most careful deliberation and planning efforts, the future may not align with assumptions made about it. This process is aimed at offering decision teams the conviction that they have done their best to make as good a decision as possible, all things considered.

. . . .
Q Why should we use this process? What are the benefits?

A The current work climate is often harried. We are pushed to make decisions quickly. In addition, we live and work with people who are different than we are – who see the world differently than we do. And we operate in hierarchical contexts where there are significant differences in power – and having more power doesn't always mean being more wise!

These factors together can lead to decision-making that fails to take into account the best evidence, that is unclear about what should be most important within the context of the work, and that inappropriately privileges the views of some perspectives over others.

This process is designed to gain trust, to achieve well-considered, quality decisions, and to effectively implement such decisions. As the views of relevant individuals are considered in the decision process, decisions are legitimized and are more likely to achieve compliance, even when people disagree with them.

"The ethics consult we had using this process was extremely helpful. It helped the team decide what the important issues were and how we wanted to prioritize our concerns. It also assisted us with an action plan based on our priorities. It gave us a framework to assist with deciding on a plan of action. Using the process, Bashir was able to assist staff in working through the issues systematically and dispassionately rather than basing decisions on emotional reactivity. Staff members were able to drill down and articulate the ethical issues behind the reactivity."

Barbara Friesen
Clinical Coordinator South Burnaby Mental Health Centre

Facilitating Conversations

It is highly recommended that someone be given the responsibility to lead the decision team through the process. The facilitator can be a member of the team or someone from outside the group.

There are several benefits to having a team member facilitate the process:

1 Assuming a trusted member is chosen, she or he will already have an established rapport with the team

2 No additional human resources will be required to undertake the process

3 Capacity will be built within the team for using the process for future issues

4 Capacity could be built for attending to the ethics dimensions of system-level issues within the day-to-day processes of the team's operations

Having a facilitator from outside the team guide the process can also be very beneficial:

1 A well-chosen external facilitator will hopefully have good facilitation skills enabling the conversation to flow in line with the process values the approach is based on

2 All team members will be able to act as participants in the process and will be free to provide content expertise as appropriate

3 If the external facilitator has experience with such processes, the team will find it easier to go through

Where organizations have access to a facilitator with a background in ethics, this can also bring important advantages:

1 If an ethicist who is familiar with the process plays this role, he or she may enable the team to go through the process more efficiently

2 An ethicist will likely be best able to effectively facilitate the discussions about values in the process

All things considered, we recommend that teams choose a skilled facilitator to guide the process. We recommend the team choose someone who:

- is able to treat people with respect (per page 21)

- is willing to use the process

- is willing to understand the group's context

- is not committed to any solution, but is open minded and willing to live up to the process values the approach is based on

- is committed to assisting the group arrive at a practical solution within a reasonable time frame, and

- is well-regarded by colleagues

A good place to start is your organization's departments for ethics services, organizational development or strategic change management.

Facilitating Conversations

. . . .

If it is decided that a team member will facilitate the process, the facilitator would be well-advised to:

1 Review the toolkit carefully, identifying any parts of the process that are unclear

2 Make a careful plan and timeline for what parts of the process she or he hopes to complete, and by when

3 Begin imagining the kinds of conversations that will occur for each of the steps

4 Pay special attention to steps 1 and 2 of the process and discuss each with the team leaders before the process begins

5 Test out the pdf forms in the worksheets to make sure they work well

6 Consider contacting the author Bashir Jiwani for additional advice as appropriate

. . . .

If it is decided that the decision team leader will facilitate the process, then we recommend the following precautions:

1 The facilitator should be able to be a neutral party in the process. Often those responsible for decisions have strong vested interests that should be understood and engaged carefully. These individuals often also occupy higher positions in the hierarchy of the organization relative to other members of the decision team.

2 If the facilitator is also the Decision Team Leader, it may be very difficult for them to remain neutral in the conversation. And having to remain neutral may not create sufficient space for their own concerns and perspectives to emerge. Even if these perspectives do emerge in a clean way, their relative position in the organizational hierarchy may prevent honest engagement of this perspective.

For these reasons, it may be more effective to separate these roles if resources permit. If this is not possible, the facilitator should take steps to try to minimize the impact of these worries. Mitigation strategies could include:

1 Naming this conflict of interest as a reality with the decision team

2 Making space within the conversation for group members to check in on how the facilitation is working

3 Creating a mechanism for the group members to provide anonymous feedback on how well the facilitation is working

4 Carefully reviewing the elements to good facilitation section in this book (p. 15) and/or consulting other resources on how to facilitate conversations well

Facilitating Conversations

"This decision map provided a framework for our interprofessional team to problem solve ethical challenges in our work place. This systematic approach assisted us to understand and then support resident requests. These requests are often complicated with complex family dynamics and the mapping process minimized the complexities. Following the map, we identified barriers and could take necessary steps to remove them. This document assisted our team to develop healthy dialogue throughout the process. We found the mapping process kept the resident in the center of the planning and decision making, by providing feedback loops to maintain our focus."

Glenda Wonnacott
Manager, Operations Residential Care and Assisted Living

ELEMENTS OF GOOD FACILITATION

Effectively using the process in this guide requires enabling the shared work team to use the skills of active listening in a pluralistic space. Facilitators will have to:

- Create trust within the decision group
- Clarify and interrogate the facts in context within the group
- Build a shared understanding of what matters to different individuals, and a shared commitment to what should matter most
- Explain each step the team goes through so participants understand what they're doing and why

Group facilitation is an area of work that comes with its own skill set, knowledge base, and character traits. In addition to these, for a facilitator to be successful in this process for system-level decision-making, they will need to be skilled at:

- Demonstrating and helping others demonstrate respect
- Asking good questions
- Reframing responses
- Distinguishing between facts, values and emotions
- Acknowledging participants' emotions and their perspectives about facts and values and helping to collectively work through these

....

1 Demonstrating respect:

- Treating others with unconditional positive regard
  - › "Regardless of what I think of your opinion, I will treat you well"
- Listening empathetically to understand
  - › "I will work hard to open my mind to understand why you think what you think"
  - › "I will work hard to open my heart to understand what you are feeling"
- Engaging others' ideas
  - › "I will share some of my thinking with the hope of sharpening/ deepening each of our perspectives and moving to a broader view"
- Supporting the team to treat each other with respect
  - › "We commit to conversations where each of us treats each other gently, kindly and well, listens to understand before arguing, and showing up and offering our honest, considered perspectives"

....

2 Asking good questions:

- Open ended
  - › Allowing the teller's story to emerge
- Probing
  - › Exploring the reasons behind perspectives

3 **Reframing responses:**
- To demonstrate understanding
  - ⟩ So participants are heard and can believe in the process
  - ⟩ So the best thinking informs the discussion
- To neutralize
  - ⟩ Making hard to hear perspectives more palatable
  - ⟩ Enabling connections between participants
- To deepen understanding
  - ⟩ By helping tellers hear and reflect on their own starting points of view

4 **Distinguishing between the facts, values, and emotions. Assisting others to understand and make these distinctions:**
- Facts
  - ⟩ What is descriptively true in the context
  - ⟩ What is real or people perceive to be real (as opposed to how we want things to be)
- Values
  - ⟩ What is important about the issue
- Emotions
  - ⟩ How one is feeling, in heart and body, about the situation

5 **Acknowledging facts, values, and emotions in a teller's story without judgement. Helping individuals or groups work through these:**
- Facts
  - ⟩ "Your understanding is that... (e.g. if we do *x*, then *y* will happen)"
  - ⟩ "You believe that... (e.g. *x* people will need *y* resources to achieve *z*)"
- Values
  - ⟩ "From your perspective, we should be most worried about... (e.g. keeping people safe)"
  - ⟩ "In your view it's really important that... (e.g. we don't waste time or energy)"
- Emotions
  - ⟩ "You're feeling... (e.g. annoyed, angry, scared)"

The next few pages offer tips for facilitating the different elements of the process. Additional tips can be found within each Step.

HELPING A GROUP GET THE FACTS STRAIGHT

Facts concern the descriptive backdrop of the story: a description of the world as it is. Who is involved? What is happening? What is the context?

....

Examples of good questions for understanding facts:

Open Ended:

- What is your understanding of the situation?
- How did we arrive at this situation?
- What else do you need to know?
- What else would you like to know?
- If this situation continues unchanged how do you think it will affect:
  › You?
  › Your team?
  › Your other colleagues/the organization?
  › Patients/families?
  › Society at large?

Probing:

- How did you come to this understanding?
- What examples are you thinking of that lead you to this concern?
- What makes you think so?
- What evidence are you relying on?

HELPING A GROUP ARTICULATE VALUES

Values concern what's at stake in the story. They are often arrived at by examining how we want the world to look differently from how it does today. What matters here? Why?

....

Examples of good questions for understanding values:

Open ended:

- What is important to you as we move forward?
- Why do you prefer this solution – what does it give you that you believe is important?
- What would a solution have to achieve for you to be satisfied with it?
- In our society, this value (e.g. equity) is important – what is your sense of this? What do you think this value means in our context? How important is it?
- If someone who would disagree with your perspective were here, what would they say is important? How would you respond?

Probing:

- Why is this important?
- Here is a competing value (tell story); how would you balance these two?
- What would have to happen for you to change your mind? What does this tell you about what else matters to you?

SUPPORTING EMOTIONS

How are people feeling? What is going on in their hearts and bodies (as opposed to in their minds)?

. . . .

Examples of good language for debriefing emotions:

Exploring:

- What is in your heart as you go through this?
- How are you feeling about this?
- You seem very...
- Are you feeling....?

Acknowledging:

- This must be very difficult for you.
- I'm sorry you have to go through this.

It is often difficult for people to discuss how they are feeling. Instead of referring to our emotional state we often use the word "feeling" to describe what we think about an issue.

Being familiar with different descriptions of emotions can help clarify what a feeling is and open a conversation about emotions. The following page lists a number of words that describe emotional or feeling states. The facilitator may benefit from becoming familiar with some of these to be better able to help the participants describe how they are feeling in or about the situation under discussion.

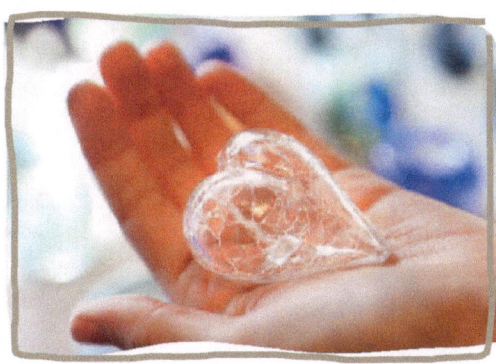

WORDS DESCRIBING DIFFERENT EMOTIONS

adored	defeated	good	mad	sensitive
afraid	dejected	greedy	mean	shaky
aggravated	delighted	grief-stricken	melancholic	shocked
agitated	depressed	groovy	mischievous	shy
agonized	desired	grouchy	miserable	silly
alarmed	desirous	grumpy	moody	sleepy
alienated	disappointed	guilty	mortified	sorry
amazed	discouraged	happy	neglected	spiteful
amused	disgusted	hassled	nervous	stressed
angry	disliked	hateful	numb	surprised
anguished	dismayed	helpless	optimistic	suspicious
annoyed	distressed	hesitant	ornery	sympathetic
antsy	disturbed	homesick	outraged	tender
anxious	down	hopeful	overwhelmed	tense
apprehensive	dreadful	hopeless	panicked	terrified
aroused	eager	horrible	passionate	thrilled
ashamed	ecstatic	hostile	patient	tired
astonished	edgy	humiliated	peaceful	tormented
at ease	elated	hurt	pessimistic	triumphant
attracted	embarrassed	hysterical	pleased	troubled
awful	empathetic	impatient	proud	uncomfortable
awkward	encouraged	indifferent	puzzled	uneasy
bashful	enraged	infatuated	queasy	unhappy
bewildered	enthralled	inferior	rageful	unsafe
bitter	enthused	insecure	raptured	unsettled
blissful	envious	insulted	regretful	upset
bored	euphoric	irate	rejected	vengeful
brave	exasperated	irked	relieved	vicious
calm	excited	irritated	reluctant	victorious
caring	exhausted	isolated	remorseful	warm
cautious	exhilarated	jealous	resentful	weary
cheerful	fatigued	jittery	restless	woeful
comfortable	fearful	jolly	revulsed	wonderful
compassionate	ferocious	joyous	ridiculous	worried
concerned	fidgety	lazy	riled	wrathful
confident	frightened	leery	rushed	yucky
confused	frustrated	liked	sad	zany
contempt	funny	loathed	safe	zeal less
content	furious	loathing	satisfied	zestless
critical	glad	lonely	scared	
curious	gleeful	loved	scornful	
cynical	gloomy	loving	secure	

Step 1 Establish the Team

Here we get clear on team makeup, mandate and work process.

1 Circulate the worksheet to be completed (on the next page) to team members in advance and ask them to prepare to discuss the questions in this step.

2 With your decision team, answer the following questions:

- Who else should be part of the team doing this work?
- Does the team commit to treating each other with respect? (see definition in sidebar).
- To whom is the team accountable?
- To whom does the team report?
- What outcome is the team looking for?
- How will the team deal with difference?
- How will the team resolve disagreement?
- Does the group commit to engaging those with relevant expertise and those affected by the decision in the process?

3 Complete the form on the next page.

4 Review Step 2 (Communication) and complete the first part of the worksheet.

WHO SHOULD PARTICIPATE ON THE DECISION TEAM?

In deciding who should be involved on the decision team, it is important that...

- The group should have the authority to settle the issue at hand (that is, to make the decision/recommendation)

- The group should have representation from the key areas affected by the decision

- The group should be small enough to ensure that each member can be treated with respect as defined below

- There is a clear administrative leader on the team with the appropriate authority to undertake this work

TREATING SOMEONE WITH RESPECT REQUIRES

- Treating them with kindness...
  > As if they have as much or more power in the relationship relative to you (even if they don't)
  > Regardless of whether or not you agree with their words or deeds

- Opening your heart and mind...
  > To try to understand and feel their standpoint

- Engaging their ideas...
  > Sharing your reasons, and working together to develop a broader perspective

TIPS FOR SUCCESS

- Although these questions may look daunting, recognize that if they are not resolved up front, the group may struggle

- Check in regularly to ensure these commitments are being followed

- For some of the questions in the worksheet (such as to whom will the group present its findings?), team members should be able to ask the team lead for guidance

WORKSHEET (Fill out accordingly)

Who are the members of the decision team?	Names:	Positions:
Who is the administrative leader on the team?	Name:	Position:
To whom will the team present its findings?	Name:	Position:
What outcome is the team mandated to achieve?	☐ Decision ☐ Recommendation ☐ Issue Analysis (with no recommendation or decision) ☐ Other:	
If there is disagreement in the group, how is this disagreement to be handled?	☐ Strong Consensus (The group will deliberate until everyone is in general agreement about the actual decision.) ☐ Procedural Consensus **(Recommended)** (The group will deliberate until the point at which, while there is not agreement on the actual decision, everyone is in general agreement that the various different perspectives have been heard, and reasons have been exchanged such that the process has been fair and the decision reached is reasonable.) ☐ Majority Rule (The group will deliberate until a majority of participants (measured by vote) agrees on the actual decision.) ☐ Stamina Rule (The group will deliberate until the most passionate and vocal member outlasts the rest of the group.) ☐ Leader Rule (The leader of the group will make the final decision.) ☐ Other:	
Do group members agree to these commitments?	• The group's attitude towards different and sometimes difficult to hear or unpopular perspectives on issues will be to actively seek these out, and try to understand and respond to them.	☐ Yes ☐ No
	• The group will engage affected stakeholders by sharing the decision team's recommendations and rationale, and by inviting and responding to feedback.	☐ Yes ☐ No
	• Group members will treat each other with kindness, seek to understand each other's perspectives, and engage together to debate reasons respectfully.	☐ Yes ☐ No
	• The group will defend beliefs about facts with evidence.	☐ Yes ☐ No

Step 2 Communication Strategy

COMMUNICATE FROM THE START

Communication should be considered from the beginning of the process.

People will need to know about and understand decisions if they are to be implemented effectively and contribute to the organization and community it serves.

This step ensures that a careful communication strategy accompanies the group's decision.

1 Determine who is affected by the issue and the decision.

2 Determine what messages need to be conveyed to these people about:

- The fact that the issue is being addressed
- The process that will be or has been used to address it
- Opportunities for providing input about evidence and about what is important in reflecting on the issue (e.g. the content from steps 8 and 14)
- The decision made, including the content and rationale behind it, and how to provide feedback

3 Revisit the communication strategy, even briefly, after every meeting

4 Complete the worksheet on Page 24 for each phase of the decision process.

TIPS FOR SUCCESS

- The communication plan should span the time frame of the decision process — from when the issue is under review to when a final decision has been reached.
- Communication should be two-way; when information is shared, a feedback mechanism should also be in place.
- There should be meaningful transparency about:
  › The fact that an issue is being considered
  › The assumptions made about the context
  › What decision leaders consider is most important in addressing the issue
  › Why a preliminary or final decision is justified
  › How those impacted can provide the decision-makers with feedback and how that feedback will be used

© The Author(s), under exclusive license to Springer Nature Switzerland AG 2021
B. Jiwani, *Good Organizational Decisions*, SpringerBriefs in Ethics,
https://doi.org/10.1007/978-3-030-33401-7_2

WORKSHEET (Fill out accordingly)

	Specific Audience	Key messages	Communication vehicle	Contact point for those wishing to follow up	Person responsible	Timeline
Before						
During						
After						

Step **2 Communication Strategy**

Notes:

_____

_____

_____

_____

_____

_____

_____

_____

_____

_____

_____

_____

_____

_____

_____

_____

_____

_____

_____

_____

_____

Step 3 — Select the Key Question(s)

SAMPLE QUESTIONS

- "What should our staff attendance policy be during a public health emergency?" (Instead of "should we force staff to come to work?")

- "Where should we allocate these scarce resources?" (As opposed to "Should we only give these resources to children?")

- "How should we respond to people who request resources but do not use them appropriately to maximize the benefit from them?" (Instead of "How do we deal with non-compliant users?")

QUESTIONS AND SUB-QUESTIONS

Some questions are actually made up of discrete sub-questions, all of which must be addressed to successfully answer the key question.

For example, any question about resource allocation will have three sub-questions:
1) What criteria should be used to compare candidates for receiving the resource?
2) Who will use the criteria and make the decision?
3) What process will be used to gather the information, analyze the options, make the decision, communicate and implement it?

In this step we define the problem and specify exactly what question we are trying to answer.

1 Determine the problem(s) the group is working to solve.

- The question you ask will determine the type and scope of answer you get

- You want to ensure that the group is working on the same problem and asking the best question to help solve that problem

- A sense of the key question will arise from discussions with the requester of the process. You want to confirm and refine this with the shared work team

2 List each suggestion as a possible question that the group might tackle.

3 For each ask, "if we get an answer to this question, will it provide sufficient direction for us to deal with the problem?"

4 Notice if there are specific smaller questions that are part of a key question and organize these together.

5 Select a key question from the list.

- Many questions will present themselves; the challenge is choosing which should be addressed in the time available

6 Do not get hung up on this step.

- The articulation of the problem may evolve with further discussion of the facts

- Get enough of a shared understanding to move forward and set the expectation you will revisit this as you move through the process

TIPS FOR SUCCESS

- Focus on a broad question that, if answered well, will spark more specific questions and provide meaningful direction for moving forward.

- Some questions are really about missing information. Park these, (and assure the group) we will get to them in the next section.

- Avoid questions that only allow yes or no answers to make possible a broad range of responses.

- Questions that begin with "What should" or "How should" work well.

- Questions that are made up of sub-questions may require separate values analyses and solutions. Set this expectation for yourself and the decision team and be prepared to adjust your process to accommodate this.

- In writing the question, only include descriptions and clauses about which there is explicit agreement.

- Try to use honest and accurate, yet morally neutral language — language that others will be able to hear without feeling threatened or judged.

- Don't worry if this step seems difficult — it is!

- Take the time to do this work well.

- Sometimes it can be helpful to engage in the facts conversation for a few minutes, then come to this step, and then return back to the facts step.

- It can be helpful for the leader(s) to:
  - › think about key questions in advance and come to the meeting with one or two possibilities in mind
  - › imagine the future they desire, and follow the line of inquiry that will make getting there possible.

Step **3 Select the Key Question(s)**

WORKSHEET (Fill out accordingly)

Questions that need to be addressed:	The Key Question the team will focus on:
1.	
2.	Specific sub-questions that are part of the broader key question: 1) 2)
3.	3)
4.	
5.	
6.	

Step **3** **Select the Key Question(s)**

Notes:

_____

_____

_____

_____

_____

_____

_____

_____

_____

_____

_____

_____

_____

_____

_____

_____

_____

_____

_____

_____

_____

Step 4 Look at the Evidence

A FACT...

Is a belief that is true about the world. The list of facts describes the landscape against which the decision is being made.

Some beliefs about reality are good candidates for scientific confirmation. For example, the root cause of a situation or how an individual or group is being or will be impacted by a given intervention can be studied.

For these types of beliefs, the more evidence that we have for it, the more likely it is a fact. The quality of this type of belief will depend upon the evidence we have to support it.

Some beliefs, on the other hand, are matters of interpretation. For example, questions such as "What happens when we die?", "Is there or isn't there a God?", and "Are human beings responsible for their actions?" are all matters that are not very good candidates for scientific analysis. They require a different sort of evaluation.

It is not likely that disagreements about these types of beliefs can be conclusively resolved in a short space of time. Nor is it necessary to resolve this type of disagreement to find common ways of moving forward for a very large majority of system-level questions.

This step is where we examine decision team members' understanding of the facts.

1 Using the worksheet, describe:

- What we know for sure about the context
- The evidence we have to base this on
- The information that is missing, but that we can find out (and who will do this research)
- The information that is missing, that we probably cannot know

2 Develop a shared understanding of the context, including areas that may be unsettled or controversial.

3 Discuss the evidence:

- Is there agreement about the sources of evidence?
- Is there agreement about how this evidence is interpreted?

4 Ask specifically if there are any assumptions people believe are contentious or unclear and make these explicit.

5 Identify the source(s) of disagreement and explore whether consensus is possible.

TIPS FOR SUCCESS

- Ensure that you list beliefs about the world (things that are true or false – often declarative sentences) and not values (what is important to us – often imperatives – "we should...")

- Remember that reasonable people can understand what reality looks like

- People can also reasonably and legitimately interpret evidence differently

- If someone states a belief that is contentious, ask probing questions to understand the source of disagreement

- Use qualifiers (e.g. "sometimes", "in most cases") to get to a statement everyone can live with

- List disagreement about a fact as a fact itself. (e.g. there is disagreement amongst the team about best practice standards related to...)

- If the list is long and it helps with clarifying, create subheadings for the facts. Examples of possible subheadings include:

  › About end users (patients, families, students, clients)

  › About the system (relevant laws, policies, and processes)

  › About the team members involved

  › About the community

- Capture in the story parts of the current context that are working well

- It may be useful to begin to append and reference key documents into the toolkit for the consult

WORKSHEET (Fill out accordingly)

What we know for sure...	Our evidence for this is...
About the end users	
About the service providers	
About relevant laws and policies	
About the broader community	
About the system	
What we don't know but can find out...	The person responsible for getting this information...
What we don't know and would have to guess about...	

Step 5 Consider What's Important

SAMPLE WORKSHEET ENTRIES

In our solution it is important that...

- We build trust with the users of our service

- We promote independence in our clients

- We build a sense of community amongst those who have to carry out the decision

VALUES CAN BE...

- Instrumental/strategic: important because they lead to something of greater significance

- Intrinsic/inherent: important for their own sake

In this step we describe in detail exactly what is important to us in the issue and what we want to ensure our solution addresses. We move from discussing the world we currently see, to the world we want.

1 Brainstorm everything that the decision should live up to (worksheet on page 37).

- Ask people to offer full ideas in answering the Key Question established in Step 3: "Whatever our answer, it is important that...".

- List all considerations regardless of degree of importance.

2 Prioritize the list. (See techniques for prioritizing value statements on next page.)

3 Review the list and confirm the prioritization.

4 Discuss the justification for the prioritization: why is it reasonable to prioritize and balance in this way?

5 Identify values about which there was disagreement and discuss how to address these.

The resulting list will be the criteria against which the quality of different options will be judged.

APPENDIX C...

Provides a glossary of terms which may be useful for team members to review. The list is not comprehensive. Individuals and teams may also interpret some of these words differently from each other and from the definition offered. That is okay. What is important is to be clear about what specifically matters to the group and to justify its place relative to other value statements.

TIPS FOR SUCCESS

- Make explicit what matters in the situation
- In determining what the group wants more of, look at:
    - positive experiences in the context — consider why these are attractive
    - difficulties experienced currently — what matters that is not currently there
- When a consideration is identified as important, explore whether it is important for its own sake or because it gives us something else of more significance
- If the latter, be sure to capture both the instrumental and the intrinsic value on the list
    - For example, if it is important that "all team members clearly document conversations with users about what is important to them", is this important,
        - "to minimize exposure to legal liability,"
        - "to ensure consistency of care or service,"
        - "to respect the user's values and autonomy," or
        - "to assist family to understand the perspective of their loved one"?
    - All of these may be important, but some will likely be more important based on the context. So all of the considerations in quotes above should be listed.
- Avoid one-word values that are open to interpretation.

TECHNIQUES FOR PRIORITIZING VALUES

Possible ways to do this exercise include:

- **Group Scoring**: If the group is small enough, discuss each value and agree on a score of between 1 and 5 (where 5 is crucial).
- **Private Scoring**: Have each individual score each value privately on paper, then collect and tabulate.*
- **Numbers on the wall**: Write each value statement on a whiteboard or flip chart and have everyone write their score beside each, then tabulate.*
- **Colours on the wall**: Give each participant 5 colours of stickers/dots, identify each one with a score between 1 and 5, and ask each participant to place one dot beside each value statement, then tabulate.*
- **Small Group**: Ask everyone to score each statement individually, then have them work in groups to discuss their ratings and to identify: 1) a collective score for each value statement where agreement was reached, and 2) values for which no consensus was reached. Then tabulate.*

* To tabulate, add up the scores applied to each value, and divide by the number of scores given. This will give you an average score. Using the average score, list the values in descending order. This will give you the values statement in rank order.

WORKSHEET (Fill out accordingly)

However we answer the key question _____, it is important that...	Priority:				
	Important 1	Very 2	Important 3	4	Crucial 5
	☐	☐	☐	☐	☐
	☐	☐	☐	☐	☐
	☐	☐	☐	☐	☐
	☐	☐	☐	☐	☐
	☐	☐	☐	☐	☐
	☐	☐	☐	☐	☐
	☐	☐	☐	☐	☐
	☐	☐	☐	☐	☐
	☐	☐	☐	☐	☐
	☐	☐	☐	☐	☐
	☐	☐	☐	☐	☐
	☐	☐	☐	☐	☐
	☐	☐	☐	☐	☐
	☐	☐	☐	☐	☐
	☐	☐	☐	☐	☐
	☐	☐	☐	☐	☐
	☐	☐	☐	☐	☐
	☐	☐	☐	☐	☐
	☐	☐	☐	☐	☐
	☐	☐	☐	☐	☐
	☐	☐	☐	☐	☐
	☐	☐	☐	☐	☐
	☐	☐	☐	☐	☐
	☐	☐	☐	☐	☐

Step 5b Synthesize Values (optional)

THE EMERGENCE OF CORE VALUES

One of the benefits of this process is that it allows the team's most important commitments to emerge.

For teams who work together regularly, the first time the team uses this process an initial set of values will be defined. This list can become a starting point for future exercises, where the team's sense of each of these values can be solidified, and what the values mean for the group can be further specified.

The list of values will likely expand as the team explores issues with slightly different contextual features.

Over time, a set of core commitments will emerge that can guide the team when facing an issue in the heat of the moment.

The group may choose to organize the value statements that emerge in Step 5 according to themes, using the worksheet on page 40. To do this, the group should:

1 Identify a value theme to the left of the detailed value descriptions.

2 Group/sort the value statements according to value theme (see Sample 2).

3 Use the value themes as the guiding values in the table in Step 6 (see Sample 3).

TIPS FOR SUCCESS

- Recognize that the challenge with value words by themselves (such as "respect for dignity", "efficiency", etc.) is that different people can interpret these terms differently. The detailed descriptions tied to each value in Step 4 become the way that this group specifies the meaning of that term.

- As the value word is used in the future, both within the group and in external communication, be sure to clarify what the group means using the more descriptive value statements.

WORKSHEET (Fill out accordingly)

Value	Value Description (However we answer the key question _____ , it is important that...)	Average Score	Rank

Step 6 Brainstorm Options

BRAINSTORMING CAN...

- Help loosen the cognitive stranglehold that can capture a group, especially when dominated by one or a few powerful group members who have set opinions

- Enable participation from all decision team members

- Allow creative solutions to emerge

In this step, the decision team creatively explores the kinds of solutions, conventional or not, that might uphold the values prioritized in the previous step.

1 Explain the rules of brainstorming – anything goes, no judgment for 5-10 minutes.

2 Invite members to provide possible ways of answering the question.

3 List each solution on a flip chart.

4 If a proposed solution is vague, probe the participants to develop it further – not to judge it but to ensure it is developed enough that people understand it and can evaluate it.

TIPS FOR SUCCESS

- Do not evaluate solution possibilities, simply list them
- If anyone challenges a possible solution or offers critical feedback, acknowledge the challenge but don't engage it; ask that this evaluation be held until the next step
- Usually a problem has at least two extreme solutions – it can help to name these. This is also a good way to get the brainstorm started

© The Author(s), under exclusive license to Springer Nature Switzerland AG 2021
B. Jiwani, *Good Organizational Decisions*, SpringerBriefs in Ethics,
https://doi.org/10.1007/978-3-030-33401-7_7

Step **6 Brainstorm Options**

WORKSHEET (Fill out accordingly)

The Key Question for which we are brainstorming:

Possible ways of answering the question:

Step 7   Analyze Options

LIVING WITH INTEGRITY INVOLVES:

- Identifying what matters most
- Articulating and justifying why this is so
- Carefully choosing a course of action that best allows one to live up to what matters
- And then making it happen!

In this step the team looks at the possible solutions to see which ones best allow the organization and its people to live with integrity. We systematically analyze the possibilities brainstormed in Step 6 against the standards we set out in Step 5.

Facilitators may also consider using an online version of this tool available at: http://incorporatingethics.ca/public_files/exercise/ethics_analysis.htm

1 List the highest priority values (from Step 5) in the top row of the table.

- Use the list of detailed value statements or the value themes

- If the list of values or value themes is too long, make a decision about how many to use

- This decision should be informed by the separation between values in the value ranking, the importance of the question being examined, and the time available

- Aim for between seven and ten value statements

2 Choose some of the options brainstormed (from Step 6) and list them in the first column.

3 Pick an option and go through each of the prioritized values asking, "How well does this option live up to this value?"

4 Systematically discuss how well the option in question lives up to the value in question. If an option's consistency with a value depends on some other factor, name this contingency.

TIPS FOR SUCCESS

- Choose options that appeal to the group, but also one or two that at first glance seem unlikely to work

- When doing this as a group, put the list of prioritized values next to the list of possible options (flipcharts are helpful for this)

- This step can also be completed less systematically but more quickly by looking at the emergent prioritized values and asking what solution might best live up to these. Experience suggests that while this approach is faster, a full systematic analysis leads to a more justified and comprehensive solution.

© The Author(s), under exclusive license to Springer Nature Switzerland AG 2021
B. Jiwani, *Good Organizational Decisions*, SpringerBriefs in Ethics,
https://doi.org/10.1007/978-3-030-33401-7_8

Step 7 **Analyze Options**

WORKSHEET (Fill out accordingly)

The Key Question for this discussion is:

List Options Here	List Most Important Values Here					

Step 7 **Analyze Options**

Notes:

_____
_____
_____
_____
_____
_____
_____
_____
_____
_____
_____
_____
_____
_____
_____
_____
_____
_____
_____
_____
_____

Step 8 — The Preliminary Decision

BEING TRANSPARENT CAN...

- Help those impacted to better understand why the decision was made
- Make it easier for others to contribute:
  › More, different or better evidence
  › Reasons for weighing values differently
- Increase the likelihood that the decision will be carried out even if those who must do so disagree with it
- Help those affected to better understand the organization's values and their own
- Help all concerned to live with greater integrity

This step is where we articulate the decision team's thinking so far in the process. We spell out as clearly as possible our understanding of the context, what should matter most and why, and what solution we believe best lives up to these values — at this stage of the group's reasoning.

1. Build a preliminary solution that incorporates the best of all of the solution options from Step 7.

2. Articulate the preliminary decision and the reasoning behind it:
   - Describe the proposed solution
   - Name key factual assumptions that the solution turns on and that anyone assessing the solution will need to understand
   - Describe the values the decision lives up to
   - Identify any values the decision doesn't live up to
   - Provide reasons why this is the best balance

TIPS FOR SUCCESS

- In building the solution anticipate a cobbled approach where elements of various possible responses are combined together in a manner that best lives up to the values identified
- Use as simple language as possible
- Think about who might disagree with the decision and the language you will need to use to help them understand your justification
- Be sure to provide the reasons why some values are prioritized over others

© The Author(s), under exclusive license to Springer Nature Switzerland AG 2021
B. Jiwani, *Good Organizational Decisions*, SpringerBriefs in Ethics,
https://doi.org/10.1007/978-3-030-33401-7_9

WORKSHEET (Fill out accordingly)

Question...	
Salient facts...	
We recommend that...	
This allows us to best...	
This solution does not...	
We argue that this is justified because...	

Notes:

_____

_____

_____

_____

_____

_____

_____

_____

_____

_____

_____

_____

_____

_____

_____

_____

_____

_____

_____

_____

_____

Step 9 Engagement

GOOD ENGAGEMENT LEADS TO...

- Quality/Best Practice: from greater accuracy of information about the context and reasons for acting
- Compliance: from those who might disagree but are still going along because they trust a process that has taken their concerns seriously
- Legitimacy: from having given those affected by a decision the opportunity to influence it
- Trust: from those affected being provided open and honest information, and having their input genuinely sought

TREATING THE PUBLIC WITH RESPECT IN PUBLIC POLICY DECISIONS REQUIRES...

- Transparency: people should know what decisions are being made
- Inclusion: people should be able to influence these decisions
- Deliberation: influence should come from participation in a respectful conversation where reasons are exchanged
- Education: the conversation should include an opportunity for the public to understand the issues in question so they may participate in an informed way.
- Recursiveness: past decisions should be open for ongoing critical reflection and feedback.
- Reflexiveness: decisions should be based on reasons, not power, authority, or access to resources.

In this crucial step, we plan how to share our thinking with others in order to engage their thinking, receive their feedback and learn from them. We then carry this plan out.

1 Identify the different stakeholder groups affected by the issue.

- Stakeholders, either or both:
  › Have expertise about technical (including clinical) or social facts
  › Are affected by the situation and have a sense of what should matter in how it is resolved
- Stakeholders will likely include:
  › Those working on behalf of the organization. (This will include staff, volunteers and leaders. In health care it includes physicians)
  › Those served by the organization. (This includes clients and their families. In health care, it includes patients and residents)
  › Subgroups of the public

2 With the help of the worksheet on page 56, identify whether you will engage particular stakeholders, when this should happen, and who will be responsible.

3 For each stakeholder group, ask:

- Who will be engaged?
- What will be the purpose of the engagement?
- What forum or method will best serve the purpose of this engagement?
- What education will they need and how will this be provided?
- How will deliberation amongst the group be facilitated?
- How will participants' views be collected?
- How will you/the organization respond to this feedback?

SEE THE DETAILED TIPS FOR SUCCESS ON THE NEXT FEW PAGES

TIPS FOR SUCCESS

- Explore whether it is possible to institutionalize engagement strategies so this is not simply ad hoc work
- Check to see what organizational resources are available to support you
- Be sure to use the feedback form provided in Step 14
- Don't be scared by divergent perspectives — diversity is a source of strength!
- Ensure you pay attention to the process of the discussion and use skilled facilitators

Who to Engage

- Think about whose perspective should be heard, not just who is easiest to involve
- Pay special attention to individuals and groups who are relevant but would usually not be involved, and explore ways of facilitating their participation
- If the conversation in the decision-process refers to a group, they should probably be included in the engagement strategy
- If yours is a public service or institution, seriously consider engaging subgroups of the broader public

Purpose of Engagement

- Acknowledge out loud the importance of engagement:
  - › To make sure we have the most accurate understanding of the context
  - › To make sure we have considered carefully all of the things that should matter in the situation
  - › To identify solutions or solution dimensions the team may not have thought of
  - › To build trust
  - › To give decisions legitimacy.

Types of Engagement Forums

- Large group (over 25 people) sessions can be useful to reach a large number of people. However, they should:

  › Include some small group conversation to maximize discussion

  › Involve a large number of support people to monitor the understanding of the issue amongst the participants and to answer questions that participants might have

- Mid-size group (8-25 people) sessions can balance the reach of a larger group with more opportunity to participate.

  › Such meetings should still include some small group conversation to maximize participation in the deliberation

- Small groups (2-8) provide the best opportunity to engage participants in a more intimate way, however, they can be intimidating for some participants.

  › Discussion leaders should pay special attention to creating a safe environment for conversation

- Individual interviews can provide a good opportunity to explore a single person's perspective and can be the easiest to arrange. However, it can be difficult to create a space for exchanging ideas in this context.

  › Interviewers should be skilled at offering competing views and challenging participant perspectives in a safe way

- Consider naturally existing pathways.

  › It may be useful to consider how the kinds of people the team will consult with already meet, and to try to build the engagement exercise into one of these events

Step **9 Engagement**

Education

- Engagement leads to legitimacy when those who are included in the conversation are:
  - › Supported to understand the context of the decision
  - › Provided education to understand the content of the decision
- Ensure any information provided is relevant and detailed enough, without being overwhelming.
- Focus on the values at stake over the technical details.
- Remember people have different learning styles so present information in various media - orally, in writing, using presentations.
- Consider using interpreters and translated materials for non-English speaking audiences.

Deliberation

- Through the respectful exchange of ideas with others, participants:
  - › Can develop their own perspectives about what should matter in the decision
  - › Can share their perspectives about what should matter in the situation
  - › Have their ideas heard without pre-judgement
  - › Have their ideas considered against competing claims
  - › Have their ideas responded to after careful consideration

Collecting Feedback

- Ensure feedback is recorded in each session (see feedback worksheet from Step 14).
- Check with participants that what is recorded accurately reflects their concerns.
- Be clear with the participants about:
  > What will happen to the information collected and how the ideas and perspectives generated will inform the decision process
  > The final decision and how they can provide ongoing feedback

Responding to Feedback

- Create a summary feedback and response document compiled from the feedback received from all the consultations. (Consider anonymizing the feedback so the focus is on the content and not on who provided it.)
- Have a mechanism for getting back to the group with feedback.
- For each point heard by the facilitators/interviewers:
  > Summarize the perspective
  > Describe how the insight was reflected in the decision or provide reasons why the insight was not agreed with, if it was not reflected in the final decision
- Share the feedback and response document with all those who provided feedback.
- Ensure the forum for sharing is reasonably accessible by the target audience.

Step **9 Engagement**

WORKSHEET (Fill out accordingly)

We commit to engaging (details on ensuing pages)	Broad Timeline	Person Responsible
☐ Clients		
☐ Families/Loved Ones		
☐ Staff		
☐ Physicians		
☐ Subgroups of the Public		
☐ Others:		

. . . .

SERVICE RECIPIENTS CAN PROVIDE:

- Context about how they are affected by the system
- Outsiders' views of system operations
- Personal information about their own values — what matters to them and what causes them distress

Why engage users of the service?

- The purpose of civil society and government institutions (including in the health system) is to serve the needs of members of the community. These individuals are the experts about their own needs.
- Service recipients understand best the impact of system decisions on users of the system.
- Service recipients are representatives of the community and can provide insight into community values.
- In health care especially, patients are the experts about their goals of care and systems should be designed to best meet these goals.

WORKSHEET (Fill out accordingly)

Specific Audience	Mode of Engagement	Steps to Create Safety in the Discussion	Mode of Obtaining Feedback	Mode of Feedback Response	Engagement Session Lead	Timeline

....

**FAMILY MEMBERS &
LOVED ONES OFFER:**

- Context about how their loved ones are affected by the system
- Contextual information about how they are affected by the system
- Outsiders' views of system operations

Why engage families & loved ones?

- They are often directly impacted by system decisions so their wellbeing and personal integrity is at stake.

- They understand well the impact of system decisions on direct users.

- They can be close to users and are often able to give voice to user perspectives. This is especially true in health care where patients and residents may have their ability to express their own perspectives compromised by illness or injury.

- They are representatives of the community and can provide insight into community values.

WORKSHEET (Fill out accordingly)

Specific Audience	Mode of Engagement	Steps to Create Safety in the Discussion	Mode of Obtaining Feedback	Mode of Feedback Response	Engagement Session Lead	Timeline

. . . .

STAFF CAN PROVIDE:

- Technical information about the issue.
- Context about relational dynamics and system function.
- Personal information about their own values — what matters to them and what causes them angst.

Why engage staff?

- They likely have key information that is needed for a good decision.

- They often have to discharge the decision, so their professional and personal integrity is directly on the line.

- They are at the point of service/care and need to be able to explain decisions to users (including patients and their families).

- Users of a service build relationships with the people providing the service or care and so usually have most trust in these individuals. (This is especially true in health care; patients trust care providers.) So care and service providers will likely have a central role in the successful implementation of the decision.

WORKSHEET (Fill out accordingly)

Specific Audience	Mode of Engagement	Steps to Create Safety in the Discussion	Mode of Obtaining Feedback	Mode of Feedback Response	Engagement Session Lead	Timeline

Step **9 Engagement**

. . . .

PHYSICIANS CAN PROVIDE:

- Technical information about the issue.
- Context about relational dynamics and system function.
- Personal information about their own values — what matters to them and what causes them angst.

Why engage physicians?

- They have expertise that is needed for a good decision.
- They have to discharge the decision, so their professional and personal integrity is directly on the line.
- They need to be able to explain decisions to patients and loved ones.
- Patients trust physicians, so physicians are key to successful implementation of the decision.

WORKSHEET (Fill out accordingly)

Specific Audience	Mode of Engagement	Steps to Create Safety in the Discussion	Mode of Obtaining Feedback	Mode of Feedback Response	Engagement Session Lead	Timeline

SUBGROUPS OF THE BROADER PUBLIC OFFER:

- Context about the values of the different communities served by the system

Why engage subgroups of the broader public?

- Civil society and government organizations (including the health care system) are initiatives aimed at serving the public good and in most places reflect commitments to democratic ideals.

- Democratic ideals require that those impacted by decisions know about and can influence them.

- Many of these services are delivered at arm's length from government. A government is elected, its leader appoints other leaders who work with bureaucracies to appoint Boards, who hire CEOs, who hire other leaders, who guide policy. This distances decision-makers from the public – whose values decisions should reflect.

- People in the general public or in the sub-groups whose needs these institutions focus on often don't know about important policies that shape what and how services are provided.

- Public engagement can help bridge this distance.

WORKSHEET (Fill out accordingly)

Specific Audience	Mode of Engagement	Steps to Create Safety in the Discussion	Mode of Obtaining Feedback	Mode of Feedback Response	Engagement Session Lead	Timeline

Step **9 Engagement**

Notes:

Step 10 The Decision

TIPS FOR SUCCESS

- Use simple language.
- When thinking through competing perspectives about the facts, be clear about what evidence is in tension. Explain why you believe some evidence or analysis is better.
- When thinking through competing perspectives about the values, articulate your reasons for preferring certain value perspectives above others.
- Imagine the argument that people who disagree with your perspective might make. Reflect on how you would respond.
- Conduct the headline test: Ask the group: if your decision was broadcast on tomorrow's front page, would it appear supported by the best evidence and a fair and reasonable understanding of what's important?

This is where the decision team carefully reviews and evaluates feedback received, develops responses and revises the preliminary decision or recommendation to a final one.

1 Explore how the team's understanding of the facts has changed based on the feedback received. (Reference the facts worksheet and make appropriate changes.)

- Consider whether competing understandings of the facts is based on different sources of evidence or different evaluation of the same sources
- Decide on whether you agree or disagree with this new information. Articulate your rationale for agreeing or disagreeing

2 Explore how the team's understanding of what is most important has changed. (Reference the values worksheet and make appropriate changes.)

- Articulate your reasons for agreeing or disagreeing with competing understandings of how values should be balanced.

3 Explore how these changes impact the team's analysis. (Reference the decision analysis worksheet and make appropriate changes.)

4 Complete the issue articulation worksheet on the following page, which asks you to state:

   1 The key question to which this decision or recommendation responds
   2 The final recommendation or decision
   3 The values on which the decision is based
   4 Core sources of evidence
   5 Key decisions made in analyzing unclear or contentious evidence
   6 Reasons why this handling of the evidence is justified
   7 Key value interpretations or balancing that was done
   8 Why this balancing is justified

5 Create a high-level summary of the process followed to arrive at the decision. Include who the members of the decision team were and who was consulted in the process.

© The Author(s), under exclusive license to Springer Nature Switzerland AG 2021
B. Jiwani, *Good Organizational Decisions*, SpringerBriefs in Ethics,
https://doi.org/10.1007/978-3-030-33401-7_11

Step **10** **The Decision**

WORKSHEET (Fill out accordingly)

For the question...	
We recommend (or decide) that...	
This recommendation best allows us to...	
We relied on these sources of evidence	
Key decisions we made in analyzing unclear or controversial evidence were	
We feel this was justified because	
Key decisions we made in balancing certain values were	
We feel this was justified because	

Step 11 — Education Plan

EDUCATION

It is likely that those affected will require additional skills or information for the decision to be carried out effectively.

Fair accountability requires that those being held accountable for performing a duty to a given standard have access to the resources needed to fulfill this responsibility. Access to education or training is one such resource.

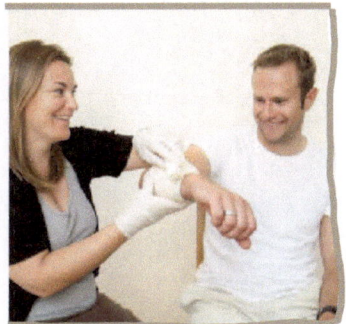

Step 11 guides education planning for those for whom the decision will mean changes in practice or process.

(Steps 11-14 cover important steps in preparation of implementation which is covered in Step 15.)

1 Ask:

- What individuals or groups will have to change their practice or process as a result of this system-level decision?
- What skills or knowledge will they require to support them in this change?
- What is the best pathway to support these individuals to acquire these skills or knowledge?
- Who will be responsible and accountable to lead this education initiative?
- How will those impacted be able to modify their training to ensure it best meets their needs?

2 Complete the worksheet on Page 66.

TIPS FOR SUCCESS

- Include those who will require education support in the consultation phase of the decision process.
- Consider:
  › The language that the target audience will need to make sense of these ideas
  › What opportunities will be required to practice using the new knowledge or skills
  › What vehicle might best meet these needs
  › The realistic outcomes of such vehicles
  › How the target audience might be informed about and attracted to participate in the education
- Think about the organization over time:
  › How new teams and individuals will be provided education
  › How this education will be updated and maintained for existing individuals and teams
  › Have groups been exposed to previous policy changes and education in the past?

© The Author(s), under exclusive license to Springer Nature Switzerland AG 2021
B. Jiwani, *Good Organizational Decisions*, SpringerBriefs in Ethics,
https://doi.org/10.1007/978-3-030-33401-7_12

Step **11** **Education Plan**

WORKSHEET (Fill out accordingly)

Specific Audience	Objectives	Education Vehicle(s)	Sustainability Plan	Lead	Timeline

Notes:

Step 12 — Downstream Support Plan

CHANGE AND MORAL COMPROMISE

When the values underpinning the system-decision conflict with what matters to those who must carry it out, it can leave affected individuals in a morally compromised situation.

WHAT AFFECTS MY INTEGRITY?

My ability to live with integrity is impacted by:

- My own skills and abilities

- My courage to live up to my values and beliefs

- The decisions made in my organization

- The support in my organization for dealing with moral distress

In this step we anticipate whose integrity might be compromised by the decision and identify strategies to mitigate this harm.

1 Answer the following questions:

- Can we reasonably anticipate any individuals or teams who must carry out the decision to disagree with it? If so, what will likely be the cause of the disagreement?

- If disagreement is about what is important (values), are good reasons being offered by the decision team to justify the approach taken?

- What ethics support resources might those affected benefit from?

- How can those who disagree with the decision appeal?

2 Complete the worksheet with the appropriate information.

TIPS FOR SUCCESS

- The decision team has an obligation to anticipate compromise to individual integrity due to upstream decisions and to assist those affected to maintain their integrity as much as possible

- Clearly indicate the values on which the system decision is based (following previous steps in this guide)

- Ensure the affected group has been included in the consultation phase and their reasons have been heard and responded to

- Include the worksheet from Step 14 in the policy, with directions for how and where to submit it and what to expect in return

- Ensure those who will be held responsible for providing education have the necessary resources to do so

© The Author(s), under exclusive license to Springer Nature Switzerland AG 2021
B. Jiwani, *Good Organizational Decisions*, SpringerBriefs in Ethics,
https://doi.org/10.1007/978-3-030-33401-7_13

WORKSHEET (Fill out accordingly)

Who will be affected by the decision?	Is the group's integrity likely to be compromised by the decision?	Have the group's perspectives been considered in the decision process?	What would assist those affected?	What support can be provided in this direction?	Who will lead the downstream support initiative for this group?

Step **12** **Downstream Support Plan**

Notes:

EVALUATION AND SUSTAINABILITY

- **Testing assumptions.** System-level decisions are usually based on a wide variety of assumptions. If the assumptions are not tested and corrected for where they were mistaken, then the goals of the decision may not be met.

- **The longer term.** System-level change usually requires intensive resources. For this investment to be maximized, the change should be linked to long-term sustainability strategies.

In this step the decision team develops a plan to ensure that the change called for remains relevant and justified over time. (This planning step is also to be implemented in Step 15.)

1 Develop an evaluation strategy for the overall change and each of the change elements.

2 Assign responsibility for evaluation.

3 Create a timeline for:

- When evaluation will be undertaken
- When feedback will be reviewed and analyzed by those leading the change

4 As part of the evaluation ask:

- Did the factual assumptions that the decision is based on remain accurate, or did the world turn out to be different somehow?

- What new information (scientific or social data) is available that might affect the approach taken?

- How effectively did the decision solve the problem?

- How well did meeting the objectives lead to achieving the desired outcomes?

- How well did the change created by the solution allow the organization to live up to the relevant values?

- Were the values considered really what mattered, or did other concerns (not initially considered) turn out to be relevant?

- What additional communication would have been useful?

- What additional education would have been useful?

- What additional downstream support would help those affected to live with integrity?

- Have we responded adequately to those who have provided feedback on the issue?

- Are we still open to feedback on the decision?

© The Author(s), under exclusive license to Springer Nature Switzerland AG 2021
B. Jiwani, *Good Organizational Decisions*, SpringerBriefs in Ethics,
https://doi.org/10.1007/978-3-030-33401-7_14

- How well did our process engage our team and stakeholders?

- How have relationships or patterns of interaction changed as a result of our intervention?

- What, if any, historical tensions or system barriers have gotten in our way and how might those be overcome?

TIPS FOR SUCCESS

- Remember a good decision does not guarantee a good outcome

- Keep the evaluation and sustainability plans in view as the team goes through the decision process

- Ensure appropriate resources for evaluation and sustainability are built into the implementation budget

- Consult external resources for additional evaluation tools as appropriate

- Consider:
  › both qualitative and quantitative data gathering strategies

  › how important it is that this decision is lived in the organization in the middle and long term

  › how much turnover is likely to be in the organization

  › who (individual or office) is best able to review the decision over time

WORKSHEET (Fill out accordingly)

Broad Change Being Implemented					
Overall Change Sponsor/ Leader					

Evaluation component	Leader	Evaluation data to be gathered by date	Evaluation data to be reviewed and analyzed by date	Changes to approach to be made by date	Resources Required

Step **13** **Evaluation and Sustainability Plan**

Notes:

Step 14. Ongoing Feedback Plan

KEEPING THE CONVERSATION GOING

Continuing the conversation is another way to ensure the best understanding of values and facts are available. It can also be an effective way to support the integrity of those affected by the decision.

This step allows the organization to stay connected with those who wish to provide feedback on the decision. (This is implemented as part of Step 15.)

Use the worksheet on the next page to:

1 Develop a plan to make the decisions and their justifications available to appropriate audiences.

2 Prepare an ongoing feedback instrument (such as the form later in this section).

3 Develop a plan to collect and respond to feedback.

4 Develop a mechanism to review the decision and adjust the implementation strategies in response to feedback.

TIPS FOR SUCCESS

- Be clear about the type of feedback people can provide
- Ensure that those who can provide feedback have access to the decision and its rationale
- Assist those providing feedback to be specific about what dimension of the issue they are concerned about
- Provide information about who to contact with questions about the process
- Clearly communicate how people's feedback will be used
- Consider keeping a policy repository on the organization's website with access to the feedback form located there as well

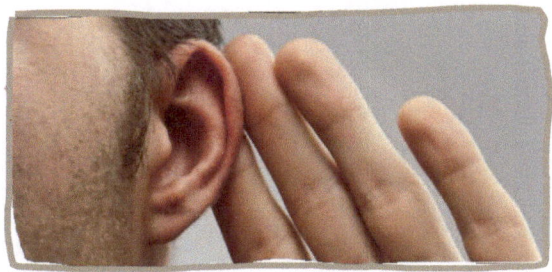

WORKSHEET (Fill out accordingly)

Who can provide feedback on the decision?	
How will they know that this is a possibility and where can they obtain a feedback form?	
Who will be responsible for collecting this feedback?	
Who will be responsible for responding to this feedback?	
When/how often will the group meet to revisit the decision based on feedback obtained?	
Who will be responsible for revising the implementation plans as appropriate?	

Your Feedback is Important!

If you have any critical feedback or suggestions for improvement we ask that you complete the following form and submit to _____
This can be done by _____

Your feedback will be acknowledged within _____ and_____will provide you with a response to your reflections within_____day(s).

THANK YOU FOR TAKING THE TIME TO HELP US DEVELOP THE MOST JUSTIFIED RESPONSE TO THIS QUESTION!

What do you like about this policy and/or the process used to generate it?	
What is your concern with the policy?	
If you think we are missing an important fact, or haven't got something right, tell us what we need to know.	
If the disagreement concerns the values, what is important that is not appropriately accounted for in the decision?	
If you believe an alternate solution would better live up to these facts and values, please share it.	
If the disagreement concerns process, what process elements should have happened differently? (e.g. Is there an individual or group that should have been consulted but was excluded?)	

Step **14 Ongoing Feedback Plan**

Notes:

Step 15 — Implement the Decision

IMPLEMENTATION

A decision by itself likely won't solve the problem. The decision has to be implemented and supported.

Implementation and follow up should be treated with care and be part of the decision process.

Here we want to be clear about how the decision and the follow up plans in Steps 10 - 14 will be implemented.

1 Respond to the following questions:
  - What specifically is being implemented? What change in the process is expected?
  - Who is affected by the change?
  - What partnerships are needed?
  - Who within the leadership is sponsoring the change?
  - Who is managing what part of the change? What are they responsible for and to whom are they accountable?
  - What is the general timing of the change?
  - What resources, financial, human, or other, will be required?

2 Complete the worksheet with the appropriate information.

3 Get the approval of the appropriate authority in the organization.

4 Implement the decision and the communication, education, downstream support and ongoing feedback plans (developed in Steps 10 - 14).

© The Author(s), under exclusive license to Springer Nature Switzerland AG 2021
B. Jiwani, *Good Organizational Decisions*, SpringerBriefs in Ethics,
https://doi.org/10.1007/978-3-030-33401-7_16

TIPS FOR SUCCESS

- Recognize that the communication, education, downstream support, and ongoing feedback steps in the process are part of the broader implementation plan
- Divide implementation into the various dimensions involved and develop a step-by-step approach for each
- Specify:
  - › the phases of the project
  - › the key deliverables in each phase
  - › the major activities for each deliverable
  - › the key milestones
  - › who is responsible for each major activity.
- Consult the change management team for more detailed planning tools and resources
- Be as clear as possible about who is most appropriate to accept accountability for discharging each part of the plan
- Ensure that those given this accountability have the resources to live up to these expectations — or change the accountability

WORKSHEET (Fill out accordingly)

Broad Change Being Implemented	
Overall Change Sponsor/ Leader	

Specific Deliverable/ Element	Leader	Responsible for	Accountable to	Change Partners	Timelines	Resources Required

Step **15 Implement the Decision**

Notes:

When	What	Specific Toolkit Sections (if any)	Who
Before meeting 1	• Review process, determine the decision team, and confirm this is the process to be used	• Welcome to this Resource (page 2) • In a Nutshell (page 4) • How to Use this Resource (page 5) • What's Different About this Approach (page 6) • Frequently Asked Questions (page 9)	Decision Team Leader
	• Decide on facilitator	• How to Use this Resource (page 5) • Frequently Asked Questions (page 9)	Decision Team Leader
	• Plan the process	• How to use this resource (page 5) • Frequently Asked Questions (page 9) • Facilitating Conversations (page 12)	Facilitator
	• Anticipate possible key question(s)	• Step 3: Select the Key Question(s) (page 27)	Decision Team Leader and Facilitator
	• Share plan with decision team and ask them to think about Steps 1 & 2	• Step 1: Establish the Team (page 21) • Step 3: Select the Key Question(s) (page 27)	Facilitator
Meeting 1	• Confirm Team Deliberation Parameters • Determine Key Question • Discuss the context • Begin discussion about values • Briefly look at the communication worksheet to see who should be alerted to the fact that this work is being undertaken	• Step 1: Establish the Team (page 21) • Step 2: Communication Strategy (page 23) • Step 3: Select the Key Question(s) (page 27) • Step 4: Look at the Evidence (page 31) • Step 5: Consider What's Important (page 35)	Decision Team
After Meeting 1	• Consolidate discussion into workbook, draft meeting 2 agenda and circulate		Facilitator
	• Reflect on facts conversation and values	• Step 4: Look at the Evidence (page 31) • Step 5: Consider What's Important (page 35)	Decision Team

Appendix **A Sample Meeting Schedule**

When	What	Specific Toolkit Sections (if any)	Who
Meeting 2	• Confirm Key Question • Review work to date and make necessary changes • Confirm and prioritize values • Brainstorm options • Analyze Options • Determine Preliminary Decision • Briefly look at the communication worksheet to see who should be alerted at this stage	• Step 2: Communication Strategy (page 23) • Step 4: Look at the Evidence (page 31) • Step 5: Consider What's Important (page 35) • Step 6: Brainstorm Options (page 41) • Step 7: Analyze Options (page 43) • Step 8: The Preliminary Decision (page 47)	Decision Team
After Meeting 2	• Consolidate discussion into toolkit and circulate		Facilitator
	• Reflect on experts and affected parties whose perspectives should inform the decision	• Step 9: Engagement (page 51)	Decision Team
Meeting 3	• Review work to date and make necessary changes • Develop engagement plan	• Step 9: Engagement (page 51)	Decision Team
After Meeting 3	• Consolidate discussion into toolkit and circulate		Facilitator
	• Undertake engagement plan	• Step 9: Engagement (page 51)	Per engagement plan
	• Collect engagement feedback, consolidate, and circulate to the decision team • Review consolidated feedback and reflect on implications	• Step 10: The Decision (page 63)	Facilitator Decision Team
Meeting 4	• Share findings • Discuss implications of findings • Make final decision • Develop Communication Plan	• Step 2: Communication Strategy (page 23) • Step 10: The Decision (page 63)	Decision Team

Appendix A Sample Meeting Schedule

When	What	Specific Toolkit Sections (if any)	Who
After Meeting 4	• Consolidate discussion into toolkit and circulate		Facilitator
	• Undertake communication next steps • Reflect on education, downstream support, evaluation and sustainability, ongoing feedback, and implementation plans	• Step 11: Education Plan (page 65) • Step 12: Downstream Support Plan (page 69) • Step 13: Evaluation & Sustainability Plan (page 73) • Step 14: Ongoing Feedback Plan (page 77) • Step 15: Implement the Decision (page 81)	Decision Team
Meeting 5	• Review work to date and make necessary changes • Complete education, downstream support, evaluation and sustainability, ongoing feedback, and implementation plans	• Step 11: Education Plan (page 65) • Step 12: Downstream Support Plan (page 69) • Step 13: Evaluation & Sustainability Plan (page 73) • Step 14: Ongoing Feedback Plan (page 77) • Step 15: Implement the Decision (page 81)	Decision Team
After Meeting 5	• Implement		As per implementation plan

B
Important Values to Consider

Acceptability

The extent to which a solution meets the expectations of a given individual or group. Use of this value requires being clear about whose standards or expectations are of concern. This value doesn't necessarily align with ethical justification — a solution may be acceptable to an individual or group, but may still not live up to important values of others or broadly accepted social norms.

Accountability

Answerability of an individual or group to another individual or group for the kind and quality of decisions made or actions taken. Accountability can be to a supervisor or sponsor organization (e.g. to a professional association as part of one's professional mandate). It can also be to an individual or group to whom one has a fiduciary responsibility (e.g. as a group's representative at a meeting or to a community of people one accepts a duty to serve.)

Affordability

The presence of a quantity of a resource (usually money) within available resources that is required to obtain a good or service. Sometimes affordability implies the ability to pay for something without going into debt. In principle, debt can be also be affordable if resources are available over time to pay for the extra resources required to cover the debt. The comfort or discomfort of going into some sort of debt to be able to pay for something itself depends on other values and is not inherent in the idea of affordability.

Care for the Vulnerable

The commitment to identify, prioritize, and use resources to meet the needs of those who are vulnerable.   An individual or group is vulnerable when they have limited ability or resources to defend against risks to potential harms to which they are exposed. This value is usually concerned with real risks to serious harms that arise due to factors beyond the control of the exposed individuals and groups. Those who are marginalized in some relevant respect have a greater claim on resources (time, attention, capital, human resource hours, care and concern) than those who do not face these risks.

Civil Society (Commitment to)

Civil society refers to interactions outside government where efforts are taken to advance the common good. The commitment to effective civil society reflects efforts to improve the well-being of society, often driven by volunteers, beyond formal government action.

Civil society organizations can be large and small and include faith-based and non-faith-based

© The Author(s), under exclusive license to Springer Nature Switzerland AG 2021
B. Jiwani, *Good Organizational Decisions*, SpringerBriefs in Ethics,
https://doi.org/10.1007/978-3-030-33401-7

institutions whose primary goal is to improve the quality of life of people and communities. These interactions may be inspired by different values systems and are most effective when they recognize the importance of and seek to strengthen connectedness and solidarity.

Collaboration

Two or more parties working together with the aim of understanding their own and each other's needs and building common responses to common problems in a manner that meets the needs of all parties without sacrifice.

Common Good

The understanding that humanity is somehow connected and the well-being of a community is distinct, separate from, and more than the well-being of each individual constituent.

Community

Recognizing the interconnectedness of all those who serve and are served by institutions, the commitment to nurturing the relationships through which services are planned, delivered, and received. It acknowledges that without interrelationships, service and care planners, and providers would not have work and/or would be deprived of the opportunity to serve their calling, and service and care recipients would not receive what they need to flourish in times of need or challenges to their well-being. It is about taking proactive organization-level action to strengthen the relationships of all the people in the community where the service is being provided.

Compassion

Can be understood as empathy plus action. Compassion involves opening one's mind to try to understand what the other thinks is true about the world and is important in life. It also requires opening one's heart to try to feel what it must be like to experience the world as the other. And then, if the other is suffering, it involves doing something to assist the other to cope with or remedy that suffering.

Compliance with Law

Similar to compliance with policy, the commitment to follow obligations set out in law, whether in statute, legal precedent or government regulation. People feel compelled to obey the law because of the severity of the penalties for not following, because the process of being taken to court is taxing, and because the law seems a justified reflection of the values of the society represented. These are all strong reasons. However, it is important to remember that laws are not always consistent, laws may not always be sensitive to the complexities of the context, and there may be laws that members of that society believe are not ethically justified. As with policy, if following a law goes against key values then following it would not necessarily be ethically justified.

Compliance with Policy

The commitment to understand and make decisions in line with established guides, such as institutional policy. Organizations are often bound by various directives, including professional codes of ethics, organizational values statements and organizational policies. These documents offer direction on what institutions and organizations believe are the appropriate values that should guide conduct. While these guides are a very important resource when doing an ethics analysis, they do not always offer consistent direction and are not always sensitive to the context of specific situations. It is important to remember that the main justification for following a policy is that the policy itself is based on justified values. If following a policy leads to action that goes against key values, then following it would not be ethically justified.

Compromise

Two or more parties seeking a response to a problem where each has to sacrifice something of importance in order to allow the other(s) to be able to live with the solution.

Consensus

Two or more parties coming to agreement about the best response to an issue or problem. This is different from having a shared understanding (where everyone achieves the same perception of the issues, but may disagree on solutions) or an agreement to proceed (where everyone in a group agrees to move forward in a certain direction, but may not agree that the solution chosen is in fact the best or most appropriate).

Consistency

The alignment of different actions, people, organizations, etc. For example, when two institutions use the same principles to determine the approach to compensating their staff, they can be described as consistent in these matters. Consistency can apply to the ends that are being pursued — moral resources such as values, principles, goals and objectives. They could also be the means for achieving the end — strategies and tactics.

Creating Moral Space

Ensuring that an appropriate forum is available for a sufficient period of time to consider whether a decision aligns with what is most important. Creating a moral space requires a commitment to interactions which respect human dignity (see below), where participants hear each other and are meaningfully heard.

....

Democracy

Decisions that determine the terms of a collective's association are legitimate to the extent that those affected by the decisions are able to influence them, either directly or through some mediated mechanism.

....

Duty to Care

Especially relevant in healthcare, this is the responsibility of healthcare workers to provide care to members of the community, even when this involves exposure to some risk of harm on the part of the health care provider.

....

Effectiveness

The extent to which an intervention meets the ends — the outcomes and impacts — it is designed to achieve. This requires having a clear and shared understanding of what a given intervention strategy is trying to achieve. In many cases, outcomes and impacts can be hard to measure which sometimes leads to focusing on more easily counted and evaluated interventions and outputs. When the measurable is prioritized over the important, this risks ineffective interventions, unmet ends and wasted resources.

....

Efficiency

Achieving a desired objective using the fewest possible resources – thereby with the least waste of resource. Efficiency is never an end in itself, and always a means to achieve some greater value.

....

Empathy

Understanding the world from another's point of view. This is a necessary correlate of having unconditional positive regard for others because making an effort to understand and appreciate another's views involves (and some might argue is what leads to) accepting that they are an individual who, like all of us, has developed their views based on particular circumstances that they have experienced.

....

Equal Outcome

When resources are distributed to ensure that everyone achieves the same level or state. In education, this might mean distributing resources in such a way as to ensure everyone achieves the same degree of understanding or ability. In health care this might mean everyone achieves the same status of health and wellbeing. Note that this would likely require unequal distribution of resources (as those with greater need would require more resources to achieve the outcome); the outcomes may not be the best achievable (as the goal is everyone being at the same level, regardless of the quality of that level).

....

Equality of Access

When a resource is distributed in such a way that everyone has the same access to the resource according to their need (as opposed to other criteria such as proximity to the resource, ability to pay, political authority, social status.)

Equality of Opportunity

When all members of society have an equal chance at receiving necessary resources. The focus is on the equal disbursement of the chance at receiving the resource — not the actual equal distribution of the resource or the equal achievement of the goals the resource is meant to provide. Note that this may not lead to equal outcomes.

Equity

When a resource is distributed based on need as opposed to other criteria such as ability to pay, social status, etc. Equity can mean equal distribution if everyone has the same need or unequal distribution if the needs of some are greater than the needs of others. How need is defined and measured will depend on the objectives the resource is meant to achieve.

Excellence

Performing an action or delivering a service to the highest standard — or maximally achieving the objectives of a strategy. This includes how the action is done and the outcome it achieves. Closely aligned with the concept of best-practice, it requires constantly improving quality of interventions to better meet the needs of those receiving services. It implies growth and capacity building of human resources which requires ongoing learning, research and training.

Fairness

Concerns the way people are treated in the distribution of a resource. It can refer to the process by which an allocation decision is made (procedural justice – see below). It can also refer to the pattern of distribution, or who gets what (distributive justice – see below). Within these two types of fairness, different ideas abound about what each of these types of fairness require. Because this term can be interpreted in multiple ways, it is very important to define when used in an argument.

Fidelity to Trust

The special relationship that certain individuals, usually professionals, have to those who are vulnerable and who put their trust in the professional to assist them in some manner. In health care, patients are much less powerful than healthcare professionals: They are usually ill, whereas caregivers are not; they are usually not in surroundings they are comfortable in, whereas health care professionals are; and most of all, they lack the expert knowledge and resources required to meet their healthcare goals - which health care professionals have. Patients must put themselves in the hands of their health care providers and trust that the care providers will help them to achieve the patient's goals. This puts health care professionals in a conflict of interest where the conflict lies between meeting their personal interests and the needs of their patients. Fidelity to trust is about maintaining the trust patients put in their care providers to advance patient interests ahead of their own.

Generosity

The sharing of one's time, energy, money, or other valuable resource when one is not obligated to do so. Especially giving without expecting anything in return.

Good Governance

Decisions for an organization are made through justified processes and substantively represents the values of the appropriate group over time. In the context of civil society organizations, the values tend to focus on the public good. In public health care and education systems, this includes en-ensuring that decisions are based on the best information available and on values that the public being represented sees as appropriate. This implies a transparency of decision-making and accountability to the law, to political leaders, and to the public broadly.

Honesty and Truth-Telling

Telling the truth to a party by providing full access to relevant and wanted information. Honesty relates to the value of respect for human dignity. If part of what gives human beings dignity is the ability to make sense of reality and make decisions accordingly, withholding information or providing information that is not true deprives the receiving party of exercising the very ability that is the source of their dignity.

Human Dignity

No matter one's perspective, religious or cultural background, or socio- economic status, one's life is valuable and deserves to be treated with compassion and care. This value calls for special attention to vulnerable members of the community who are easily marginalized and whose humanity is easily forgotten. It calls for three levels of response. First it requires treating others with kindness and care. Second, it requires empathy - actively listening to others with a view to understanding their minds and hearts without judgement. And third, it calls for engaging others in respectful dialogue, sharing one's own experience and perspective with a view to arriving at a broader perspective on the world.

Inclusiveness

Commitment to ensuring that room is created for all relevant people and perspectives to be involved in a decision or action. It requires identifying and overcoming barriers to participation that would prevent such participation unjustifiably. Barriers can include language, power, financial resources, time, access to decision corridors, prejudice, and even style of discussion and debate.

Individual Liberty

An individual's basic rights to freedom of speech, movement, association, etc. This value calls for free individual choice to govern matters from relationships to resource allocation. It argues against centralized intervention by external parties beyond the enforcement of contracts. On this view, individuals should be free to make agreements to exchange goods and services based on their means and preferences. Any forced participation is unjustified and individuals should not be required to enter agreements not of their choosing.

Integrity

The wholeness of an individual or group that occurs when there is alignment between the individual's or group's understanding of what matters in life (their talk) and their decisions and actions (their walk). This requires ensuring that all the decisions made and actions taken pay explicit attention to and are in keeping with the values of the individual or group – recognizing that value trade-offs will be required from time to time. Integrity is never fully realized because the life journey of every individual or group is characterized by new encounters which require reassessment of what should matter most against the new circumstances. Integrity by itself does not indicate what one's talk should be. It requires skills of moral reasoning exercised in relationship with others where the honest assessment of values and consistency between values and behaviour can be explored through dialogue.

Intellectual Honesty

Being truthful about the extent and limits of an individual or group's knowledge or expertise.

Justice - Distributive

Those who are in similar circumstances are treated in similar ways, unless there is good reason to treat them differently. Justice requires identifying what criteria are relevant when discriminating between different alternatives and ensuring these criteria are well-justified. Accounts of justice in different moral traditions prioritize different values. Approaches that focus on equity and social justice accept a special duty to those who are most vulnerable in society – those without resources necessary for living healthy, peaceful lives. Other accounts focus on values such as respect for individual freedom, maximizing the aggregate wellbeing of people in the community, remedying past injustice, and maximizing solidarity and community wellbeing over time.

Justice - Procedural(Fairness)

When decisions made in a given situation meet certain standards. These can include:

- Transparency: people should know that decisions are being made and what these are; they should have access to the criteria being used to make decisions
- Inclusion: those affected should be able to influence these decisions and not be limited by unjustified barriers
- Deliberation: influence should come from participation in a respectful conversation where reasons are exchanged
- Education: the conversation should follow an opportunity to understand the issues in question.
- Recursiveness: past decisions should be open for ongoing critical reflection and feedback
- Reflexiveness: decisions should be based on reasons, not power, authority, or access to resources

Kindness

Treating people gently and well. A useful heuristic is to think about kindness in relation to power – kindness requires treating people who have less power their relationship with you as if they have as much or more power than you.

Non-Abandonment

Commitment to provide care for an individual or group, irrespective of any criteria they might or might not meet. In health care this value is of particular concern to patients in the end-of-life context. It requires that whatever decisions are made and steps are taken in the care and treatment plan, they involve assurance and demonstration to the patient and family that even though aggressive treatments may not pursued, care for and attention to the patient will always be provided.

Patient Autonomy

Related to the ideas of self-determination and consent, a human being's dignity is partially derived from their ability to think about and make sense of their lives and to make decisions for themselves. Respecting this capacity for autonomy or self-regulation requires allowing individuals to make choices for themselves about decisions that impact them exclusively, and ensuring individuals have a say in decisions that impact them as part of a group. The superficial version of this value involves simply following the preferences of individuals in choices that concern them. The deep version of this value involves helping individuals understand the context well, think through what should matter most from their perspective, analyze options to see which best lives up these prioritized values, and then respecting the resulting decision.

In health care, this value calls for respecting the genuinely held values and beliefs of patients and having decisions about the care of the patient be guided by these values and beliefs. According to this principle, patient values and beliefs should guide decision-making about a patient's care even if the resulting decision is inconsistent with what care providers or family members may have chosen.

Based on this value, if someone was able to make meaning of their lives, their view of the world should survive their ability to actively participate in decision-making. Respecting this is one way to continue to honour and respect the dignity of what is now a particularly vulnerable human being. when a patient is not able to participate in decision-making about their care or treatment plan, an appropriate substitute decision-making process should be put in place.

Patient Benefit

The purpose of health care systems is to advance the wellbeing of those coming forward to receive care and treatment. Health care providers have the professional mandate to benefit and promote the wellbeing of their patients. This can be a challenging value to uphold for patients whose capacity to participate in decisions is compromised and in cases where what matters to patients is in conflict with the values of family members and/or members of the care team.

This value requires working to advance the wellbeing of the patient and meet their goals of care. In the culture of western medicine, wellbeing has traditionally focused on the physiological functioning of the body. But as society becomes increasingly diverse, there is growing understanding that the psychological, emotional, spiritual, and relational needs of patients should also inform discussions of patient benefit. Wellbeing and benefit should be assessed from the patient's perspective.   This relates to the values of Human Dignity and Autonomy.

Patient-Centered Care

A philosophy of practice that sees the patient's needs and perspective at the heart of the healthcare encounter. It includes a commitment to designing systems and processes to meet the needs of patients first and before the needs of care providers and the system. For example, were a treatment delivery system (say kidney dialysis) to be patient-centered, it would be provided where it is most convenient and effective for patients. For some patients, this would include home-based dialysis, where they could avoid travel, and receive treatment in a familiar environment.

Pluralism

A response to diversity that is committed to respecting difference while at the same time pursuing the possibility of shared responses to common problems. It sees difference as valuable and focuses on the search for a cosmopolitan ethic - values for guiding decisions that impact a diverse group of people that the group can agree with. Pluralism requires education about the nature and types of differences that exist within a group and then engagement where individuals and sub-groups can come together in respectful dialogue to explore the possibility of guiding values that are shared amongst the individuals and sub-groups.

Privacy (also known as Confidentiality)

Not interfering in the personal space of individuals or groups without their consent or just cause. This sees information about service recipients as part of their personal space and requires that this information not be shared without their consent.

Professional Competence

Having and using the technical expertise, including knowledge and skill base, to carry out professional roles at accepted standards of practice, based in the context of the values of the service recipient, the institution, and the organization. In the health care context and a team approach to health care delivery, part of professionally competent care is ensuring the team has a shared understanding of the goals of care for a patient and ensuring access to care and treatment that the patient is eligible for at the appropriate standard.

Professional Integrity

The alignment between a professional's values and their actions. Professionals (and service providers more generally) are moral agents who have their own values and beliefs about how to pursue a meaningful life. These values and beliefs can include personal convictions as well as professional ones, such as those found in professional codes of ethics and standards of practice.

Quality Improvement and Patient Safety

Related to excellence, the commitment to meeting the objectives of an intervention to increasingly higher standards over time. It starts with a commitment not to create new harms to patients — going against the objectives of the intervention. It requires identification and careful study of practice standards to develop increasingly better means of protecting against possible harms and improving service standards. Note that patient safety is often narrowly construed as risk to physical well-being, but should extend to holistic well-being, including psychological, emotional, and spiritual dimensions.

Wherever possible service providers should not be forced to participate in service plans they find excessively morally compromising, at least when it is possible to involve other providers who are more comfortable with the service plan.

Reciprocity

The responsibility of society (or agents of society, such as a health authority or education board) to respond to the needs and personal circumstances of staff who put themselves in harm's way in times of crisis — commensurate with the degree and conditions of the harm. In healthcare this is especially relevant in the context of various kinds of emergency situations.

Responsible Stewardship

The careful management and distribution of community resources. It is about taking good care of the resources, both human and material, that have been invested with the organization. It is not only about resource allocation, but also about careful and deliberate modeling and direction setting for how resources should be established and used to serve and advance the interests of the broader community. It involves using resources effectively to meet the intended goals with minimal waste, allocating resources in a manner that is fair to all those who need them and that is sensitive to broader questions of justice in society. Stewardship also requires ensuring the long-term sustainability of resources and clearly accounting for decisions made about the use of the resources, including the value judgments on which resource decisions are based.

Social Utility

The aggregate harms and benefits of an action for a population or group. Fair distribution strategies are ones that result in the best outcome for all. To allocate resources, we should look at all of the distribution patterns available to us, add up the good that results from each – where the consequences for each person are weighed and where every individual can get a maximum of one unit of good - subtract the harm that results from each, and then select the option that results in the most net good. The goal is to look for the strategy whose consequences allow for the greatest overall good for the greatest number. According to this view, the lives of all presently existing and future human beings are equally morally valuable in that everyone's happiness is equally morally important.

Social utility can be a challenge to calculate because it is difficult to know in advance what the consequences of a given pattern of distribution will be. Benefit and harm are themselves value-laden notions such that it is difficult to make objective calculations. It is also difficult to weigh the relative benefit and harm that results from saving the life of one person while allowing another to die. This approach can lead to some counter-intuitive scenarios. For example, if two patterns of distribution result in equal outcomes then there is no moral difference between the two on the social utility approach, even if the most vulnerable in one arrangement are treated poorly, while in another their lot is significantly improved.

....

Solidarity

Linked to the values of common good, relationships and community, this value sees the unity of the group as inherently worthwhile and important to pursue in a course of action. It favours approaches where members of the group stay together and support one another — in times of wellness and difficulty alike.

....

Spirituality

Nurturing the human spirit of all people, service providers, service recipients, and their families alike. This value is about attending to that intangible dimension of the human being, often called the soul. For many, the spirit is the seat of creativity, compassion, love, inspiration and solace. It is where we experience our connection to other members of the community. Most would agree that having a meaningful life requires paying attention to one's spirituality, however this is understood.

....

Staff and Service Provider Wellbeing

Aligned with respect for professional integrity, this value places importance on the quality of life of the service provider. It calls for minimizing risk of harm, broadly understood (physical, moral, emotional, psycho-social, etc.), to service providers and extends to taking measures to advancing their wellbeing. This includes supporting all those involved with service provision (direct providers and those in support and leadership roles) to help make the work experience an integral part of a meaningful life. It recognizes that service providers in public institutions such as health care and education have a tremendous stake in the experience of the service. Providers often spend more time with colleagues than families and derive much self-worth from their vocations. This value indicates it is important for the needs of the care giving community to be attended to by the organization.

....

Support for Relationships

Supporting people's relationships, particularly with loved ones and family members. Especially relevant in health care, this value calls for special attention to supporting the relationships that a patient has as part of the broader effort of trying to advance their health and wellbeing. The value takes on particular significance in the end of life context where patients may experience loneliness and fear of abandonment. Support for relationships in this context includes the creation of appropriate space, time and assistance to make the dying experience for patients as healthy and positive as possible.

....

....

Sustainability

Concerns the viability of policies, decisions, arrangements, or a system over a long period of time. It calls for decisions to ensure atwhatever is being sustained is not threatened and that it will be able to continue to meet its objectives into the future. It requires being clear about what the entity's objectives are, and what might threaten its ability to meet these objectives. It also requires clarity within an organization about whether the desired sustainability is about a specific strategy, service, program, or the organization itself.

....

Transparency (also known as Publicity)

Exposing the process and rationale of decision-making to affected others for viewing and   comment in a full, accurate and timely manner. It is a necessary part of demonstrating respect to affected others in the context of system-level decision-making. Poor transparency practices can lead to poor decisions, unfair suffering by those impacted, reinforcement of bad decision-making practices, poor understanding of the system, lack of trust for leadership, and missed opportunities for moral growth and reform for the community.

....

Trusting relationships

A commitment to create relationships where parties can be counted on to act consistently according to justified values. Central to any trusting relationship is honesty, open communication and transparency. To trust someone, one needs to know that they will be treated with respect. They will not be lied to or deceived and that one will be forthcoming with important, relevant information in a manner that is respectful to the relationship.

....

Unconditional Positive Regard

A fundamental element of respect for human dignity and related to kindness, this value accepts that, whatever the differences of class or culture, education, or opinion, all people are basically equal as human beings, and deserving of recognition and respect. In practical terms it means separating people from the positions they take on issues. It calls for treating individuals with kindness and gentleness, honour and dignity whatever we may think of their views. Whether or not I agree with your views, indeed even if I find them offensive or absurd, I will always treat you well. Put another way it calls for us to treat others with the same deference and kindness that is usually reserved for those who have more power in relationships than we do.

....

Acknowledgements

This work has benefited from the support of a wide variety of institutions and people. While I cannot name them all, I do wish to name and thank Fraser Health Authority and Providence Health Care in British Columbia, and the Provincial Health Ethics Network of Alberta for creating the space for this work. I am especially grateful for the support of Susan Rink, Duncan Steele, Mojisola Adurogbangba, Katherine Duthie, Al-Noor Nenshi Nathoo, and Sarah Gebauer for their involvement in the use of these materials. I also wish to acknowledge all those who have been involved in its implementation in various system-level consultations which have been the source of much learning.

I am grateful to Susan and Katherine, along with Elina Hill, for their great support in editing the book.

I'd like to say a special thanks to Dr. Michael McDonald, a mentor of mine and a pioneer in the development of such ethics-based decision-making tools. It is from Dr. McDonald's work that I draw the wisdom of the humble goal of seeking in a situation the best decision, all things considered.

This work has been developed through doctoral research that was guided by Dr. David Kahane, Dr. Colin Soskolne, and Dr. Olive Triska of the University of Alberta. I am indebted to all three for their wisdom and guidance.

© The Author(s), under exclusive license to Springer Nature Switzerland AG 2021
B. Jiwani, *Good Organizational Decisions*, SpringerBriefs in Ethics,
https://doi.org/10.1007/978-3-030-33401-7

"We were looking for a neutral approach to a problem that was fraught with emotion and complexity. What we found in using this decision-making model was that our diverse group recognized their commonly-held beliefs and consequently quickly engaged in values–based dialogues. Because we were grounded in what really mattered and because the model leads you through sorting out the social contract of the planned change, we have been able to sustain this change while remaining committed to it."

**Carla Kraft, Manager,
Public Health Delta, White Rock/South Surrey**

Bashir Jiwani, PhD